LINEAR REGRESSION ANALYSIS:
Assumptions and Applications

John P. Hoffmann and Kevin Shafer

NASW PRESS
National Association of Social Workers
Washington, DC 20002-4241

Cheryl Y. Bradley, *Publisher*
Stella Donovan, *Acquisitions Editor*
Sarah Lowman, *Project Manager*
Julie Palmer-Hoffman, *Copyeditor*
Juanita R. Doswell, *Proofreader*
Bernice Eisen, *Indexer*

Cover by Eye to Eye Design
Interior design and composition by MidAtlantic Publishing Services
Printed and bound by Hamilton Printing Company

First impression: April 2015

Library of Congress Cataloging-in-Publication Data

Hoffmann, John P. (John Patrick), 1962-
Linear regression analysis : assumptions and applications / John P. Hoffmann and Kevin Shafer.
pages cm
Includes bibliographical references and index.
ISBN 978-0-87101-457-3—ISBN 0-87101-457-2 1. Social service—Statistical methods. 2. Social service—Research—Methodology. 3. Regression analysis. 4. Linear models (Statistics) I. Shafer, Kevin. II. Title.
HV29.H64 2015
519.5′36—dc23

2014041003

Printed in the United States of America

To Lynn M. Hoffmann, a remarkable woman

—JPH

To my wife Melissa and our wonderful children

—KS

CONTENTS

ABOUT THE AUTHORS

John P. Hoffmann, PhD, is a professor in the Department of Sociology at Brigham Young University, where he teaches courses on linear regression, generalized linear models, and research methods. He received a BS in political science from James Madison University, an MS in justice studies from American University, a PhD in criminology from SUNY-Albany, and an MPH with emphases in epidemiology and behavioral science from Emory University. He is the author of several books and articles that address applied statistical methods, juvenile delinquency, and adolescent health behaviors. His current research interests include trends in adolescent and adult substance use, innovative approaches to the measurement of adolescent substance use, and the effects of stress on adolescent behaviors. He is married to Lynn Hoffmann, and they are the parents of four children: Brian, Christopher, Brandon, and Curtis.

Kevin Shafer, PhD, is an assistant professor of social work at Brigham Young University in Provo, Utah, where he teaches MSW-level courses on research methods, statistics, and community organization. Originally from Columbus, Ohio, he received a PhD in 2009, an MA in 2005, and a BA in 2002 from The Ohio State University. His research addresses stepfamily functioning, fathering in the United States and Brazil, and men's mental health domestically and abroad. He is the author of numerous research articles on these topics, which have appeared in journals such as *Social Work,* the *Journal of Marriage & Family*, the *Journal of Family Issues,* and *Family Relations.* He is married to Melissa Randall Shafer, and together they have four children—three boys and one girl.

PREFACE

We live in a world where we are surrounded by data. Studies are highlighted in newspapers, magazines, on television, and online daily. We are constantly shown graphs and charts, statistics about life expectancy, crime, pollution, unemployment, life satisfaction, elections, and other phenomena. As a social worker, you will likely encounter data frequently—standardized assessment scores, research studies, and new information as you obtain continuing education units. Understanding statistics, or at least speaking intelligently about them, is practically mandatory for well-educated people and good social workers. Yet very few people have a good grasp on the strengths and weaknesses of data collection and analysis. What does it mean to say that life expectancy in the United States is 75.8? Should we trust exit polls in the latest elections? When someone claims that "taking calcium supplements is not associated with lower risk of bone fractures in elderly women," what does this actually mean? Is it really meaningful for a social work researcher to suggest that a one-year increase in education is associated with a half-point decrease on a depression scale? These questions, and many others, are common in contemporary statistical analysis.

For budding social scientists, whether they are sociologists, psychologists, or social work researchers, it is a virtual impossibility to avoid sophisticated quantitative analyses. Such research moves well beyond simple statistics such as means, medians, modes, and standard deviations. In fact, the majority of studies found in social science (including social work) journals use sophisticated statistical models designed to predict the occurrence of one variable with information about another. The most common type of model designed to make such a prediction is the linear regression model—which is the focus of this book.

One can make a quick search anywhere and find numerous books and articles discussing the linear regression model. Students are usually exposed to this model in their second statistics course for several reasons: (1) it is relatively easy to understand; (2) statistical software designed to estimate it is widely available; (3) it is a highly flexible method; and (4) it really is the backbone of virtually all statistical analysis. Despite these very positive aspects, there are some weaknesses as well—particularly related to the misuse of the model. For instance, linear regression does not work well when two or more of the explanatory variables have high correlations (chapter 9) or if one of the data points is different from all others (chapter 12). Therefore, one of the goals of this book is to provide a relatively painless overview of the assumptions researchers and statisticians make when using the linear regression model.

However, the main purpose of this book is to show the reader how to use linear regression models in studies using quantitative data. We do so by discussing (1) why regression models are used; (2) what they tell us about the relationships between two or more variables; (3) what the assumptions of the model are and how to determine whether they are satisfied; (4) what to do when assumptions are not satisfied; and (5) what to do when the outcome variable is measured using only two categories. In chapter 13, we explain that linear regression models are not good at this last point, which will lead us to discuss the logistic regression model.

The coauthors of this book have substantial experience teaching statistics and have seen some students struggle with these methods, but many more succeed. We know that having a basic foundation of elementary statistics is critical to understanding the linear regression model. As a result, the first chapter of the book covers some elementary statistics. However, this chapter is meant as a review, not as a substitute for a full class on such methods. We recommend that you review a basic statistics textbook if you are unfamiliar with mean, medians, standard deviations, *z* scores, *t* tests, correlation, or analysis of variance. We also suggest that you take some time to learn either Stata or SPSS, statistical software packages designed to carry out the analyses covered here. Although SPSS may be the more commonly used of these programs, you will notice that Stata is often more flexible for the methods we show you in this book. If you are an SPSS user, we have noted throughout the book how to use the program to analyze the data using the drop-down menu approach. SPSS also uses syntax—however, this method is not very efficient. In contrast, we highly recommend using syntax in Stata and do-files, which are simple programs of instructions for Stata. Many books and websites will introduce you to both programs, although we strongly recommend the website maintained by UCLA at www.ats.ucla.edu/stat/. One further point for Stata users is that we rely on the command line approach in this book, but we highly encourage you to write out these commands in a do-file so that you have a record of the commands used. Using Stata's log files is also strongly recommended.

The chapters follow the typical format for books on linear regression. We begin with elementary statistics, followed by a discussion of the simple linear regression model and regressions with multiple variables. Next, we will learn about goodness-of-fit, comparison of models, and dummy variables. This will be followed by a discussion of linear regression assumptions, such as multicollinearity, heteroscedasticity, autocorrelation, and influential observations. Finally, we provide a brief overview of the logistic regression model.

We hope that you, the reader, will keep one thing in mind as you read this book. Statistics are often maligned by observers. Consider, for example, the book *How to Lie with Statistics* (Huff, 1954), which supports disbelief of statistical analyses. Sometimes social workers feel that social work is more art than science. Although we believe that social work is an art, it is also a science. It is important for social work researchers and clinicians to acknowledge that knowing what the data do and do not support is important for the field. Of course, you should have a healthy dose of skepticism and use some common sense and imagination when doing statistical analysis. We hope that you will feel comfortable using your imagination and reasoning skills as you cover the material and do your own quantitative work. Statistical analysis has led to many important discoveries in medicine, social work, and other social sciences. It has also informed policy decisions. Both clinical and macro social workers should be keenly aware of the power of statistical methods while also acknowledging its weaknesses.

We use both Stata and SPSS in our presentation. In the first few chapters, we present the output from both programs to increase your comfort level with what the programs look like. In the latter chapters, we give one set of results. We list Stata commands and, in a few instances, SPSS syntax, in `Courier New font`. When we discuss using the drop-down menus in SPSS, we provide those directions by using *italicized font.* Key words and phrases are also *italicized* for emphasis, and we provide a list of key terms and concepts at the end of each chapter. We provide you with a few exercises to do on your own at the end of the chapter. We also highly encourage you follow along with the examples we provide in the text.

ACKNOWLEDGMENTS

We would like to thank our students for all their helpful input on this book. John has taught many fine students in the sociology department at Brigham Young University (BYU) and in criminal justice at the University of South Carolina who have helped us refine this text by providing both solicited and unsolicited input. Kevin's MSW students at BYU and Dr. Gordon Limb, the director of the BYU School of Social Work, encouraged him to tailor this book for a social work audience and get it published. We are so grateful to these individuals for their confidence in our work. We also had students work on this book, including Scott Baldwin, Colter Mitchell, Bryan Johnson, Karen Spence, Kristina Beacom, Amanda Cooper, and Wade Jacobsen—all of whom worked on previous iterations of this text. We are grateful to Miles Marsala, Doug Wendt, Bryce Hurst, Todd Jensen, and Garrett Pace for their assistance and suggestions on this version of the text.

CHAPTER 1 A REVIEW OF SOME ELEMENTARY STATISTICAL CONCEPTS

Chapter Objectives

This chapter reviews elementary statistical concepts that are important for understanding intermediate methods, such as linear regression.

In chapter 1, we will

- Briefly discuss descriptive statistics, such as measures of central tendency and dispersion.
- Link the normal distribution and statistical methods.
- Discuss important concepts for linear regression, such as statistical inference, the use of samples, and variable standardization.
- Reconsider inferential methods, such as the *t* test.
- Run some basic statistical analyses in SPSS and Stata.

Elementary statistics can be confusing, especially to people who are uncomfortable with numbers. Many of us were first introduced to statistics in a pre-algebra or algebra course. However, your initial introduction probably occurred in elementary school. Do you remember the first time you heard the word *mean* used to indicate the average of a set of numbers? This likely took place in some math class very early on in your education. How about graphing exercises? Do you recall being given two sets of points and being asked to plot them on graph paper? You were introduced to the x-axis and the y-axis, or the coordinate axes.

Around the same time, or perhaps a little later, you became familiar with elementary *probability*. This likely took the form of a question such as, "What is the probability of a die being thrown and landing on a five?" You first learned that you needed to count the number of possible outcomes (there are six faces on a typical die, so there are six possible outcomes). This was the denominator. Then you counted the particular outcome. This was the numerator. Putting these two counts together, you learned that the probability of the roll coming up as five is 1/6, or approximately 0.167. This latter value is known as a *proportion*. Proportions—and probabilities—must fall between zero and one. They can easily be transformed into percentages by moving the decimal place over two spaces to the right (or multiplying by 100) and placing a percentage sign next to the number. What does this mean, though? Well, one way to consider it is to say that we expect a five to come up about 16.7 percent of the time when we roll a typical die numerous times. Of course, you can probably confirm this by rolling the die many, many times. Some statisticians refer to such a view as a *frequentist* interpretation of or approach to statistics.

Probabilities are normally presented using, not surprisingly, the letter P. One way to represent the probability of a five from a roll of a die is with $P(5)$. So we may write $P(5) = 0.167$ or $P(5) = 1/6$. You might recall that some statistical tests, such as t tests (see the description later in the chapter) or analyses of variance (ANOVAs), are often accompanied by p values. As we shall learn, p values are a type of probability value used in many statistical tests.

By combining the principles of probability and elementary statistical concepts, we may develop the basic foundations for statistical analysis. In general, there are two types of statistical analyses: *descriptive* and *inferential*. The former set of methods normally is used to describe or summarize one or more variables (recall that the term *variables* is used to indicate phenomena that can take on more than one value; this contrasts with *constants*, or phenomena that take on only one value). Some common terms that you are probably familiar with are *measures of central tendency* and *measures of dispersion*. We will see several of these measures a little later. Then there are the many graphical techniques that may be used to "see" the variable. You might recognize techniques such as histograms, stem-and-leaf plots, dot plots, and box-and-whisker plots.

Inferential statistics are designed to infer or deduce something about a population from a sample. Suppose, for instance, that we are interested in determining who is likely to win the next presidential election in the United States. We will assume there are only two candidates from which to choose: Clinton and Rice. Of course, it would be enormously expensive to ask each person who is likely to vote in the next election his or her choice of president. Therefore, we may take a sample of likely voters and ask them for whom they plan to vote. Can we infer anything about the population of voters on the basis of our sample? The answer is that it depends on a number of factors. Did we collect a good sample? Were the people who responded honest? Do people change their minds as the election approaches? We do not have time to get into the many issues involved in sampling and survey responses, so we will have to assume that our sample is a good representation of the population from which it is drawn and that people are generally honest in their responses to our inquiries. Most important for our purposes is this: Inferential statistics include a set of techniques designed to help us answer questions about a population from a sample.

Another way of dividing up statistics is to compare techniques that deal with one variable from those that deal with two or more variables. Most readers of this presentation will likely be familiar with techniques designed for one variable. These include, as we shall see later, most of the descriptive statistical methods. The bulk of this presentation, at least in later chapters, concerns a technique designed for analyzing two or more variables simultaneously. A key question that motivates us is whether two or more variables are associated in some way. As the values of one variable increase, do the values of the other variable also tend to increase? Or do they decrease? In elementary statistics, students are introduced to covariances and correlations, two techniques designed to answer these questions generally. However, recall that you are not necessarily saying that one variable somehow changes another variable. Remember the maxim, *correlation does not equal causation*? We will try to avoid the term *causation* in this presentation because it involves many thorny philosophical issues (see Pearl, 2000). Nonetheless, one of our main concerns is whether one or more variables are associated with another variable in a systematic way. Determining the characteristics of this association is one of the main goals of the linear regression model that we shall learn about later.

Measures of Central Tendency and Dispersion

Means

Now that we have some background information on elementary statistics, let's learn more about the most important measures, including how they are used and how they are computed. We will begin with measures of central tendency. Suppose we have collected data on a variable such as weight in kilograms. Our intrepid researchers have carefully placed each person in the sample on a scale and recorded their body weights. To simplify things, we will assume the researchers rounded the weights to the nearest kilogram. What would be your best guess of the average weight among the sample? It is not always the best, but the most frequent measure is the *arithmetic mean*, which is computed using the following formula:

$$E[X] = \bar{x} = \sum x_i / n$$

The term on the left-hand side of the equation is $E[X]$. This is a shorthand way of saying that this is the *expected value* of the variable X. It is often used to represent the mean. To be more precise, we might also list this term as E [weight in kg], but usually, as long as it is clear that X = weight in kilograms, using $E[X]$ is sufficient. The middle term—read as *x-bar*—may also be familiar as a common symbol for the mean.

Figure 1.1: What Is a Mean?

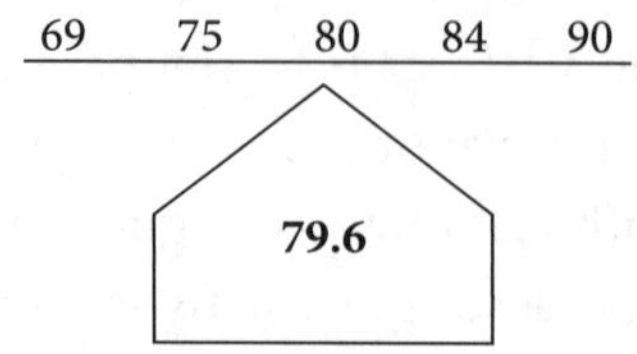

The formula for computing the mean is rather simple. We add all the values of the variable and divide this sum by the number of observations. Note that the rather cumbersome symbol that looks like an overgrown *E* in the right-hand part of the equation is the summation sign; it tells us to add whatever is next to it. The symbol x_i denotes the specific values of the x variable, or the individual weights we have measured. The subscript i indicates each observation. The symbol n represents the sample size. Sometimes the individual observations are represented as $i \ldots n$. If you know that $n = 5$, then you know there are five individual observations in your sample. In statistics, we often use uppercase roman letters to represent population values and lowercase roman letters to represent sample values. Therefore, when we say $E[X] = \bar{x}$, we are implying that our sample mean estimates the population expected value, or the population mean.

Here is a simple example: We have a sample of people's weights (in kilograms) that consist of the following set: [84, 75, 80, 69, 90]. The sum of this set is [84 + 75 + 80 + 69 + 90] = 398; therefore, the mean is 398/5 = 79.6. Another way of thinking about this mean value is that it represents the center of gravity. Suppose we have a plank of wood that is magically weightless (or of uniform weight across its span). We order the people from lightest to heaviest—trying to space them out proportionally to their weights—and ask them to sit on our plank of wood. The mean is the point of balance, or the place at which we would place a fulcrum underneath to balance the people on the plank (as illustrated in Figure 1.1).

There are some additional things you should know about the mean. First, it is measured in the same units as the observations. If your observations are not all measured in the same unit (for example, some people's weights are in kilograms, others in pounds), then the mean cannot be interpreted. Second, the mean provides a good measure of central tendency if the variable is measured continuously and is normally distributed. What do these terms mean? A variable is measured continuously—or we say the variable is *continuous*—if it can conceivably take on any real number. Of course, we usually cannot be so precise when we measure things, so it is not uncommon to round our measures to whole numbers or integers. We also often measure things using positive numbers only; it makes little sense, for instance, to try to measure a person's weight using negative numbers. The other type of variable is known as *discrete* or *categorical*; these variables

Figure 1.2: The Normal Distribution

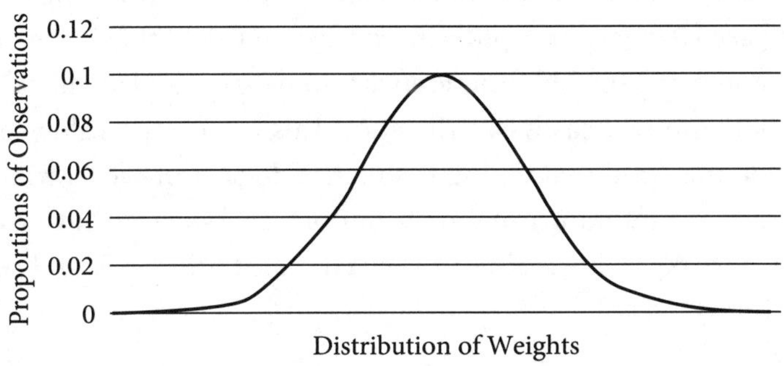

have a finite number of possible values. For example, we normally use only two categories to measure gender: female and male. Hence, this is a discrete variable.

Normally Distributed Data

We say a variable is normally distributed if it follows a bell-shaped curve. Order the values of the variable from lowest to highest and then plot them by their frequencies or the percentage of observations that have a particular value (we must assume that there are many values of our variable). We may then view the "shape" of the distribution of the variable. Figure 1.2 shows an example of a bell-shaped distribution, usually termed a *normal* or *Gaussian distribution*, using a simulated sample of weights. (It is known as Gaussian after the famous German mathematician Carl Friedrich Gauss, who purportedly discovered it.)

We will return to means and the normal distribution frequently here and throughout the book. To give you a hint of what is to come, the linear regression model is designed, in part, to predict means for particular sets of observations in the sample. For instance, if we have information on the heights of our sample members, we may wish to use this information to predict their weights. Our predictions could include predicting the mean weight of people who are 72 cm tall. We may use a linear regression model to do this.

However, suppose that our variable does not follow a normal distribution. May we still use the mean to represent the average value? The simple answer is yes, as long as the distribution does not deviate too far from the normal distribution. In many situations in the social and behavioral sciences, though, variables do not have normal distributions. A good example of this involves annual income. When we ask a sample of people about their annual incomes, we usually find that a few people earn a lot more than others. Measures of income typically provide *skewed* distributions, with long right tails. If asked to find a good measure of central tendency for income, there are several solutions available. First, we may take the *natural* (*Naperian*) *logarithm* of the variable. You might recall from an earlier mathematics course that using the natural logarithm (or, as an alternative, the base 10 logarithm) pulls in extreme values. If this is not clear, try taking your calculator and using the LN function with some large and small values (for example, 10 and 1,500). You will easily see the effect this has on the values of a variable. If you are lucky, you may find that taking the natural logarithm of a variable with a long right tail transforms it into a normal distribution. The square root or cube root of a variable may also work to "normalize" a skewed distribution. We will see examples of this in chapter 10.

Medians

Second, there are several direct measures of central tendency appropriate for skewed distributions (or other distributions plagued by extreme values such as outliers; see chapter 12), such as the *trimmed mean* and the *median*. The trimmed mean cuts off some percentage of values from the upper and lower ends of the distribution, usually 5 percent, and uses the remaining values to compute the mean value. The median should be familiar to you. It is the middle value of the distribution. To find it, we first order the values of the variable from lowest to highest. Then we choose either the middle value if there is an odd number of observations or the average of the middle two values if there is an even number of observations. If you are familiar with percentiles (or quartiles or deciles), then you might recall that the median is the 50th percentile of a distribution. The median is known as a *robust* statistic because it is relatively unaffected by extreme values. As an example, suppose we have two variables, one that follows a normal distribution (or something close to it) and another that has an extreme value:

Variable 1: [45, 50, 55, 60, 65, 70, 75]
Variable 2: [46, 51, 54, 59, 66, 71, 375]

Variable 1 has a mean of 60 and a median of 60, so we make the same estimate of its central value regardless of which measure is used (the mean and median being equal). In contrast, Variable 2 has a mean of 103 but a median of 59. Although we might debate the point, we think most people would agree that the median is a better representative of the average value than the mean for Variable 2.

Standard Deviations

The next issue to address from elementary statistics involves measures of dispersion. As the name implies, these measures consider the spread of the distribution of a variable. Most readers are familiar with the term *standard deviation*, as it is the most common measure for continuous variables. However, before seeing the formula for the standard deviation, it is useful to consider some other measures of dispersion. The most basic measure is the *sum of squares*, or SS[X]:

$$SS[X] = \Sigma(x_1 - \bar{x})^2$$

This formula first computes *deviations from the mean* $(x_i - \bar{x})$, squares each one, and adds them up. If you have learned about ANOVA models, the sum of squares should be familiar. Perhaps you even recall that there are various forms of the sum of squares. We will learn more about these in chapter 4.

A second measure of dispersion that may be more familiar to you is the *variance*, or Var[X]. It is often labeled as s^2. The formula is:

$$\text{Var}[X] = s^2 = \frac{\Sigma(x_i - \bar{x})^2}{n-1}$$

Notice that another way of computing the variance is to take the sum of squares and divide it by the sample size minus one. One of the drawbacks of the variance is that it is measured not in units of the variable but rather in squared units of the variable. To transform it into the same units as the variable, it is, therefore, a simple matter to take the square root of the variance. This measure is the standard deviation (which is why the standard deviation is noted by the letter *s*, while the variance is labeled as s^2):

$$SD[X] = s = \sqrt{\frac{\Sigma(x_i - \bar{x})^2}{n^{-1}}}$$

A variable's distribution, assuming it is normal, is often represented by its mean and standard deviation. In shorthand form, this is listed as $x \sim N(\bar{x}, s)$ (the wavy line means "distributed as"). Obviously, a variable that is measured in the same units as another and that shares the same mean is less dispersed if its standard deviation is smaller. Although not often used, another promising measure of dispersion is the *coefficient of variation* (CV), which is computed by dividing the standard deviation by the mean of the variable $(s/\bar{x})$. It is often multiplied by 100. The CV is valuable because it shows how much a variable varies about its mean.

An important point to remember is that symbols such as *s* and s^2 are used to represent sample statistics. Greek symbols—or, as we have seen up until this point, uppercase roman letters—are often used to represent population statistics. For example, the symbol for the population mean is the Greek letter *mu* (μ), whereas the symbol for the population standard deviation is the Greek letter *sigma* (σ). However, we will see when we get into the symbology of linear regression that Greek letters are used to represent other quantities as well.

Standard Error

Another useful measure of dispersion or variability refers not to the variable directly but rather to its mean. When we compute the variance or the standard deviation, we are concerned with the spread of the distribution of the variable. Typically, we have only one sample from a population—but many, many samples are possible. Let's imagine that we did take many, many samples from a population and compute a mean for each sample. We would end up with a sample of means from the population rather than simply a sample of observations. We could then compute a mean of these means, or an overall mean, which should pretty accurately reflect—assuming we do a good job of drawing the samples—the actual mean of the population of observations. Nonetheless, these numerous means will also have a distribution. It is possible to plot these means to see whether they follow a normal distribution. In fact, an important theorem from mathematical statistics states that the sample means follow a normal distribution even if they come from a nonnormally distributed variable in the population (see chapter 3). This is a very valuable finding because it allows us to make important claims about the linear regression model. We will learn about these claims in later chapters.

Our concern here is not whether the distribution of sample means is normally distributed, at least not directly. Rather, we need to consider a measure of the dispersion of these means. Statistical theory suggests that a good measure (estimate) of dispersion is the *standard error* of the mean. Why should we use the standard error? Consider the following example. In 2012, it was estimated that 42.2 million Americans, or about 15 percent of the U.S. population, used food stamps. Imagine that we wanted to know how children and adults using food stamps were similar or different from one another in terms of physical health. It is simply not practical to get information from every family using food stamps. Instead, we need to sample from this group. If we took a sample of 10,000 people from the population of 42.2 million, there would be literally trillions of possible combinations of respondents in our data. Each possible sample would be a little different from the next one. The standard error tells us how much difference we can expect between our sample and the true population mean. Think of it as a way to account for not having the full population of food stamp recipients.

The standard error is computed using the sample standard deviation as:

$$SE(\text{mean}) = \frac{s}{\sqrt{n}}$$

Standard errors are very important in linear regression analysis. Later, we will discuss another type of standard error—known as the standard error of the slope coefficient—which we use to make inferences about the regression model.

Standardizing a Variable

One of the difficult issues when we are presented with different continuous variables is that they are rarely measured in the same units. Education is measured in years, income is measured in dollars, body weight is measured in pounds or kilograms, and food intake is measured in kilocalories. It is convenient to have a way to adjust variables so their measurement units are similar. You might recall that this is one of the purposes of z scores. Assuming that we have a normally distributed set of variables, we may transform them into z scores so they are comparable. A z-score transformation uses the following formula:

$$z \text{ score} = \frac{(x_i - \bar{x})}{s}$$

Each observation of a variable is put through this formula to yield z scores, or what are commonly known as *standardized values*. The unit of measurement for z scores is standard deviations. The mean of a set of z scores is zero, whereas its standard deviation is one.

You may remember that z scores are used to determine what percentage of a distribution falls a certain distance from the mean. For example, 95 percent of the observations from a normal distribution fall within 1.96 standard deviations of the mean. This translates into 95 percent of the observations using standardized values falling within 1.96 z scores of the mean. With a slight modification, this phenomenon is helpful when we wish to make inferences from the results of the linear regression model to the population. The plot of z scores from a normally distributed variable is known as the *standard normal distribution*. As mentioned earlier, one of the principal advantages of computing z scores is that they provide a tool for comparing variables that are measured in different units; this will come in handy as we learn about linear regression models. Of course, we must be intuitively familiar with standard deviations to be able to make these comparisons.

Covariance and Correlation

Our next task involves moving from a single variable to two variables. An important use of statistics is to consider the association or relationship between two variables. As mentioned earlier, an interesting question we might ask is whether two variables shift or change together. To give an obvious and not-very-interesting example, is it fair to say that height and weight shift together? Are taller people, on average, heavier than shorter people? The answer, again on average, is most certainly yes. In statistical language, we say that height and weight *covary* or are *correlated.* The two measures most commonly used to assess the association between two continuous variables are, not surprisingly, the *covariance* and the *correlation.* To be precise, the correlation used most often is the Pearson's product-moment correlation (there are actually many types of correlations; the type attributed to the statistician Karl Pearson is the most common).

A covariance is a measure of the joint variation of two continuous variables. In less technical terms, we claim that two variables covary when there is a high probability that large values of one are accompanied by large or small values of the other. For instance, height and weight covary because large values of one tend to accompany large values of the other in a population or in most samples. This is not a perfect association because there is clearly substantial variation in heights and weights among people. The equation for the covariance is:

$$\text{Cov}\ (x,y)=\frac{\Sigma(x_i-\bar{x})(y_i-\bar{y})}{n-1}$$

The equation computes deviations from the means of both variables, multiplies them, adds up these products for each observation, and then divides this sum by the sample size minus one. Don't forget that this implies that the *x*s and *y*s come from the same unit, whether it is a person, place, or thing.

One of the problems with the covariance is that it depends on the measurement scheme of both variables. It would be helpful to have a measure of association that did not depend on these measurement units but rather offered a way to compare various associations of different sets of variables. The correlation coefficient accomplishes this task. Among the several formulas we might use to compute the correlation, the following equations are perhaps the most intuitive:

$$\text{Corr}(x,y)=\frac{\text{Cov}(x,y)}{\sqrt{\text{Var}[x]\times\text{Var}[y]}}$$

$$\text{Corr}(x,y)=\frac{\Sigma(z_x)(z_y)}{n-1}$$

The first equation shows that the correlation is simply the covariance divided by a joint measure of variability: the variances of each variable multiplied, with the square root of this quantity representing what we might call the joint or pooled standard deviation. The second equation shows the relationship between *z* scores and correlations. We might even say, without violating too many tenets of the statistical community, that the correlation is a standardized measure of association.

A couple of interesting properties of correlations are, first, that they always fall between -1 and $+1$, with positive numbers indicating a positive association and negative numbers indicating a negative association (as one variable increases, the other tends to decrease). A correlation of zero means there is no statistical association, at least not one that can be measured assuming a straight line association, between the two variables. Second, the correlation is unchanged if we add a constant to the values of the variables or if we multiple the values by some constant number. However, these constants must have the same sign, negative or positive.

As mentioned earlier, there are several other types of correlations (or what we refer to generally as *measures of association*) in addition to Pearson's. For instance, a Spearman's correlation is based on the ranks of the values of variables rather than the actual values. Similar to the median when compared to the mean, it is less sensitive to extreme values. There are also various measures of association designed for categorical variables, such as gamma, Cramer's *V*, lambda, eta, and odds ratios. Odds ratios, in particular, are popular for estimating the association between two binary (two-category) variables. Odds ratios are discussed in much greater detail in chapter 13.

Comparing Means from Two Samples

Another important topic that we will discuss before moving into linear regression analysis involves comparing means from two distributions. Of course, we may compare many statistics from distributions, including standard deviations, correlations, and standard errors, but an important issue in applied statistics is determining whether the mean from one sample (or subsample) is different from the mean of another sample (or subsample). For example, I may wish to know whether the mean income of adults from Salt Lake City, UT, is higher than the mean income of adults from Seattle, WA. If I have good samples of adults from these two cities. I can consider a couple of things. First, I can take the difference between the means. Let's say that the average annual income among a Salt Lake City sample is \$35,000 and the average annual income among a Seattle sample is \$32,500. It appears as though the average income in Salt Lake City is higher. However, we must consider something else: We have two samples, so we must consider the possibility of sampling error. Namely, our samples likely have different means than the true population means, so we should take this into account. A *t* test is designed to consider these issues by, first, taking the difference between the two means and, second, by considering the sampling error with what is known as the *pooled standard deviation*. This provides an estimate of the overall variability in the means.

The name *t* test is used because the *t* value that results from the test follows what is termed a *Student's t distribution*. This distribution looks a lot like the normal distribution; in fact, it is almost indistinguishable when the sample size is greater than 50.

Table 1.1: Descriptive Statistics from Salt Lake City and Seattle

Statistic	Salt Lake City	Seattle
M	$35,000	$32,500
SD	$500	$750
n	50	50

At smaller sample sizes, the *t* distribution has fatter tails and is a bit flatter in the middle than the normal distribution.

As mentioned earlier, the *t* test has two components: the difference between the means and an estimate of the pooled standard deviation. The following equation shows the form the *t* test takes:

$$t = \frac{\bar{x} - \bar{y}}{s_p \times \sqrt{\frac{1}{n_1} + \frac{1}{n_2}}},$$

where

$$s_p = \sqrt{\frac{(n_1 - 1)s_1^2 + (n_2 - 1)s_2^2}{(n_1 + n_2) - 2}}$$

The s_p in the above equations is the pooled standard deviation. The *n*s are the sample sizes and the s^2 represents the variances for the two groups. A key assumption that this type of *t* test makes is that the variances are equal for the two groups represented by the means. Using our example of incomes in Salt Lake City and Seattle above, let's calculate the *t* test. Table 1.1 gives the basic descriptive statistics for each town so that we may calculate this statistic.

Initially, we can plug in the mean values into the *t* test and sample sizes for each city, as seen here:

$$t = \frac{35{,}000 - 32{,}500}{s_p \times \sqrt{\frac{1}{50} + \frac{1}{50}}}$$

Of course, we need to calculate the pooled standard deviation (s_p). We need to use the sample size (*n*) and standard deviation (s^2) for each town to calculate it.

$$s_p = \sqrt{\frac{(50 - 1)500^2 + (50 - 1)750^2}{(50 + 50) - 2}}$$

No doubt, this looks like a daunting calculation, but if we take it a piece at a time, it can be easily solved.

$$s_p = \sqrt{\frac{(49)500^2 + (49)750^2}{98}}$$

$$s_p = \sqrt{\frac{12250000 + 27562500}{98}}$$

$$s_p = 6186.87$$

We can now solve the full equation:

$$t = \frac{35{,}000 - 32{,}500}{6186.87 \times \sqrt{.002}} = \frac{2{,}500}{247.47} = 10.1$$

A t value of 10.1 indicates that there is a strong statistically significant difference between the two (we will discuss this later in the chapter) because it is larger than the critical t of 1.96. As a result, we can say that the average income in Salt Lake City is indeed higher than in Seattle. Without doing this test, we would not have known that.

Sometimes, we find that the assumption of equal variances is not true; the variances differ to a large degree between two samples. When this occurs, the following test, which is known as Welch's t test, is used:

$$t' = \frac{\bar{x} - \bar{y}}{\sqrt{\frac{\text{Var}[x]}{n_1} + \frac{\text{Var}[y]}{n_2}}}$$

Unfortunately, we must use special tables if we wish to compute this value by hand and determine the probability that there is a difference between the two means. Fortunately, though, many statistical software packages provide both types of mean comparison tests, along with another test that is designed to show whether we should use the standard t test or Welch's t test.

An important assumption that we are forced to make if we wish to use these mean-comparison procedures is that the variables follow a normal distribution. The t test, for example, does not provide accurate results if the variable from either sample does not follow a normal distribution. There are other tests, such as those designed to compare ranks or medians (for example, the Wilcoxon-Mann-Whitney test), which are appropriate for nonnormal variables.

There are many other types of comparisons that we might wish to make. Clearly, comparing two means does not exhaust our interest. Suppose we wish to compare three means, four means, or even ten means. We might, for instance, have samples of incomes from adults in Salt Lake City, Utah; Seattle, Washington; Reno, Nevada; Portland, Oregon; and Boise, Idaho. We may use ANOVA procedures to compare means that are drawn from multiple samples. Using multiple comparison procedures, we may also determine whether one of the means is significantly different from one of the others. Books that describe ANOVA techniques provide an overview of these procedures. As we will learn in subsequent chapters, we may also use linear regression analysis to compute and compare means for different groups that are part of a larger sample.

Samples, Inferences, Significance Tests, and Confidence Intervals

The last substantive topic to cover in this chapter involves the core of inferential statistics: How do we know that what we found actually reflects what is occurring in the population? The cynical—but perhaps most honest—answer is that we never know whether what we found says anything accurate about the population. After all, statistics has been called the science of uncertainty for good reason. We can offer only degrees of confidence that our results reflect characteristics of the population.

However, what do we mean by a *population*? Populations may be divided into *target* populations and *study* populations. Target populations are the group about which you wish to learn something. As this might include a group in the future ("I wish to know the risk of heart attacks at 60 years old among people who are now 40 years old"), we typically find a population that closely resembles the target population; we call this the study population. There are clearly many types of populations. For instance, we might be interested in the population of sea lions off the coast of San Diego in July 2005; the population of poodles in New York City; or the population of voters in Massachusetts during the 2014 election. Yet quite a few people, when they hear the term *population* used in statistics, assume it means the U.S. population, the world's population, or some other extremely large group.

A *sample* is a set of items selected from the population. There are many books and articles on the various types of samples that one might draw from a population. The most commonly described in elementary statistics is the *simple random sample*. This means that each item in or member of the population has an equal chance of being in the sample. There are also clustered samples, stratified samples, and many other types. Most of the classic theoretical work that has gone into inferential statistics is based on the simple random sample, however.

It should be obvious by now that one of the valuable results of statistics is the ability to say something about a population from a sample, but recall the lesson we learned when discussing the standard error of the mean: We usually take only one sample from a population, but we could conceivably draw many. Therefore, any sample statistic we compute or test we run must consider the uncertainty involved in sampling. The solution to this dilemma of uncertainty has been the frequent use of standard errors for test statistics, including the mean, the standard deviation, correlations, and medians. As we will see in chapter 2, there is also a standard error for slope coefficients in linear regression models.

These standard errors may be thought of as quantitative estimates of the uncertainty involved in test statistics. They typically are used in one of two ways. First, recall from elementary statistics that when we use a *t* test, we compare the *t* value to a table of *p* values. All else being equal, a larger *t* value equates to a smaller *p* value. This approach is known generally as *significance testing* because we wish to determine whether our results are significantly different from some other possible result. It is important to note, though, that the term *significant* does not mean *important*. Rather, it was originally used to mean that the results signified or showed something (see Salsburg, 2001). We often confuse or mislead when we claim that a significance test demonstrates that we have found something special.

Showing where *p* values come from is beyond the scope of this presentation. It is perhaps simpler to provide a general interpretation. Earlier we calculated a *t* test to compare the mean income levels in Salt Lake City ($n = 50$) and Seattle ($n = 50$). If we look up a table of *t* values (available online or in most introductory statistics textbooks), we find, using a sample size of 100, that a *t* value of 10.1 corresponds to a *p* value of less than .001. This leaves approximately .001 of the area under the curve that represents the *t* distribution. One way to interpret this *p* value is with the following long-winded statement: Assuming we took many, many samples from the population of adults in Salt Lake City and Seattle, and there was actually *no difference* in mean income from these populations, we would expect to find a difference of $2,500 or something larger (in Salt Lake City) only one time, on average, out of every 1,000 samples we drew.

If you remember using null and alternative hypotheses, such a statement may sound familiar. In fact, we can translate the above inquiry into the following hypotheses:

$$H_0\text{: Mean Income, Salt Lake City} = \text{Mean Income, Seattle}$$
$$H_a\text{: Mean Income, Salt Lake City} > \text{Mean Income, Seattle}$$

Astute readers may notice that we have set up a *one-tailed* significance test. This is important because one- and two-tailed significance tests imply different comparisons. A *two-tailed* significance test, for example, is a test of whether the mean is higher in Salt Lake City *or in* Seattle. We will return to the interpretation of *p* values in chapter 2.

The second way that standard errors are used is to compute *confidence intervals* (CIs). There are many applied statisticians who prefer CIs because they provide a range of values within which some measure is likely to fall. Here, we contrast a point estimate and an interval estimate. Means and correlations are examples of point estimates: They are single numbers computed from the sample that estimate population values. An interval estimate provides a range of values that (presumably) contains the actual population value. A CI offers a range of possible or plausible values. Those who prefer CIs argue that they provide a better representation of the uncertainty inherent in statistical analyses.

The general formula for a CI is:

$$\text{Point estimate} \pm [(\text{confidence level}) \times (\text{standard error})]$$

The confidence level represents the percentage of the time, based on a z statistic or a t statistic, you wish to be able to say that the interval includes the point estimate. For example, assume we have collected data on violent crime rates from a representative sample of 100 cities in the United States. We wish to estimate a suitable range of values for the mean violent crime rate in the population. Our sample yields a mean of 104.9 with a standard deviation of 23.1. The 95 percent CI is computed as:

$$95\%\ \text{CI} = 104.9 \pm \left[1.96 \times \left(\frac{23.1}{\sqrt{100}}\right)\right] = \{100.4, 109.4\}$$

The value of 1.96 for the confidence level comes from a table of standard normal values, or z values. It corresponds to a p value of .05 (two-tailed test). The standard error formula was presented earlier in this chapter.

How do we interpret the interval of 100.4 – 109.4? There are two ways that are generally used:

1. Given a sample mean of 104.9, we are 95 percent confident that the population mean of violent crime rates falls in the interval of 100.4 and 109.4.
2. If we were to draw many samples from the population of cities in the United States, and we claimed that the population mean fell within the interval of 100.4 – 109.4, we would be accurate about 95 percent of the time.

In subsequent chapters, we will discuss how it is also possible to construct CIs for point estimates from a linear regression model.

An Example Using SPSS

The file *GSS 2010.sav* is an SPSS data set that contains many variables from the 2010 edition of the General Social Survey. We are going to use SPSS to show some of the test statistics we have discussed in this chapter. Move the variable *mntlhlth*, which measures

Table 1.2: SPSS Output

Descriptive Statistics

	N	Mean		Std. Deviation	Variance
	Statistic	Statistic	Std. Error	Statistic	Statistic
Days of poor mental health past	1,151	3.83	.216	7.315	53.515
Valid *N* (listwise)	1,151				

Figure 1.3: Stata Output

Days of Poor Mental Health Past

	Percentiles	Smallest		
1%	0	0		
5%	0	0		
10%	0	0	obs	1151
25%	0	0	Sum of Wgt.	1151
50%	0		Mean	3.825369
		Largest	Std. Dev.	7.315374
75%	4	30		
90%	15	30	Variance	53.51469
95%	25	30	Skewness	2.426593
99%	30	30	Kurtosis	8.238027

number of poor mental health days in the past 30 days, into the variable box. Click *Options* (bottom right corner) and place a check mark in the boxes labeled Mean, Std. deviation, Variance, and S.E. mean. Then click *Continue.* After it returns you to the *Descriptive Statistics* window, click *OK.* (If you wish to see how SPSS uses syntax files, try clicking *Paste* and it will bring up a new screen with the coding.) SPSS's output screen is shown in Table 1.2.

An Example Using Stata

Using the same data in Stata format, *GSS 2010.dta,* we can run the same analyses as we did with SPSS. After opening the data set, use the summarize command with the detail option to see descriptive statistics of poor mental health days. The detail option allows you to see additional statistics, such as the median and variance. After entering the following text in Stata's command line—summarize mntlhlth, detail—the output screen in Stata can be found in Figure 1.3.

Key Terms

The following is a list of key terms and concepts in chapter 1. You can find definitions to these words throughout the chapter.

- Categorical
- Causation
- Confidence intervals
- Constants
- Correlation
- Covary/correlate
- Descriptive
- Deviations from the mean
- Discrete
- E[X] = expected value of the variable X
- Frequentist
- Inferential
- Measures of central tendency
- Measures of dispersion
- Mean (arithmetic)
- Median
- Natural logarithm
- Normal/Gaussian distribution
- One-tailed significance test
- Population
- Probability
- Proportion
- Robust statistics
- Sample
- Sigma—summation
- Significance testing
- Simple random sample
- Standard deviation
- Sum of squares
- Trimmed mean
- *t* test
- Two-tailed significance test
- Variables
- Variance
- x_i
- x-bar

Analysis Exercise

1. As an exercise, see whether you can compute the standard error of the mean from the mean and the standard deviation (abbreviated Std.) from the variance listed in the table.

   SPSS: Use *Analyze-Correlate-Bivariate* to estimate the correlation between **educ** and **mntlhlth** in the GSS 2010 data (make sure the Pearson box is checked). Can you figure out how to estimate the covariance between these two variables? (Hint: use *Options* on the *Bivariate Correlations* screen.) If you are interested in confidence intervals for the mean, try using *Analyze-Descriptive Statistics-Explore* and place Public Expenditures in the *Dependent List* box.

Stata: Use the correlate or pwcorr command to estimate the correlation between **educ** and **mntlhlth** (for example, `correlate educ mntl-hlth`). Can you figure out how to estimate the covariance between these two variables? If you are interested in confidence intervals for the mean, use the ci command.

Solution

You should find that the correlation is .748 and the covariance is 37.065. For public expenditures, you should obtain 95 percent CIs of 37.18 and 50.05. Using the GSS 2010 data, look for statistically significant differences by gender in personal income with a *t* test.

SPSS: Go to Analyze → Compare Means → Independent Samples T-test to run the analysis. Use *p income* as your test variable and gender (1 = male and 0 = female) as your grouping variable. What does this table show? What is the *t* value? What is the *p* value?

Stata: In the GSS 2010 data file, you will find a variable called *gender*, which is coded as 0 = male and 1 = female (use the command codebook gender to see some information about this variable). We will use it to compare personal income (labeled *pincome*) for males and females. The command `t test pincome, by(gender)` should produce a table with which to accomplish this task. What does this table show? What is the *t* value? What is the *p* value? Unlike SPSS, you can try using the subcommand welch to request Welch's version of the *t* test for unequal variances (*t* test pincome, by(gender) welch). What does this test show?

CHAPTER 2 SIMPLE LINEAR REGRESSION

Chapter Objectives

This chapter covers simple linear regression analysis, which is often called bivariate regression. It provides an overview of the model before moving onto regression models with greater complexity.

In chapter 2, we will

- Discuss what linear regression analysis is.
- Provide an overview of the regression line and its probabilistic nature.
- Address how to run linear regression in SPSS and Stata.
- Cover how to interpret linear regression results like coefficients, p values, and other helpful statistics.
- Examine hypothesis testing in regression.

As mentioned in the Preface, a general goal of regression analysis is to estimate the association between one or more *independent* variables and a single *dependent* variable. As the names imply, a dependent variable is presumed to depend or be systematically predicted by the independent variables. However, the independent variables are thought to independently affect the dependent variable. We shall see in later chapters that using the term *independent* is somewhat misleading because these variables may, if the model is set up correctly, depend on one another. Perhaps this is why many researchers prefer to use the names *predictor* and *outcome variables*, *explanatory* and *response variables*, *exogenous* and *endogenous variables*, or some other terms. The outcome, response, or endogenous variable is synonymous with the dependent variable, while explanatory, response, and exogenous mean the same thing as independent variable.

In general, the goals of a regression analysis are to predict or explain differences in values of the dependent variable with information about the values of the independent variables. There are four things that we are interested in from a regression analysis:

1. The form of the relationship among the dependent and independent variables, or what the equation that represents the relationship looks like.

2. The direction and strength of the relationships, which are based on the value and size of the slope coefficients.
3. Which independent variables are important and which are not. This issue is based on comparing the size of the slope coefficients (which has some problems) and on the p values (or confidence intervals [CIs]).
4. Predicting a value or set of values of the dependent variable for a given set of values of the independent variables.

We will begin by considering an elementary situation: the simple linear regression model, or what some people refer to as *bivariate regression*. The term *simple* is not used to suggest that a research question is uninteresting or crude but rather that we have only one independent and one dependent variable. However, you might ask, if there are only two variables, why not use the Pearson's correlation coefficient to estimate the relationship? This is certainly a reasonable step; we can learn something about the relationship between two variables by considering the correlation (assuming they are continuous and normally distributed). However, the difference here is primarily conceptual: We think, perhaps because of a particular theory or conceptual model, that one variable influences another, or we have control over one variable and therefore wish to see its effect on another (for example, we control the amount of fertilizer and see how a particular species of tomato plant grows).

The dependent variable is usually labeled as Y, and the independent variable is labeled X (for now, let's assume we are interested in population values). To help you understand the regression model, it might be helpful to think about your pre-algebra or algebra class when you were asked to plot points on two axes. You probably labeled the vertical axis as y and the horizontal axis as x. You then used an equation such as $\mathrm{Y} = m\mathrm{X} + b$ to either represent the systematic way the points fell on the graph, or to decide where to put points on the graph. The equation provided a kind of map that told you where to place objects. The m represented the slope and the b represented the intercept, or the point at which the line crossed the y-axis. If you can recall this exercise in graphing, then simple linear regression is easy to understand. Consider Figure 2.1: What is its slope and intercept? Because we do not know the units of measurement, we cannot tell. However, we do know that the slope is a positive number and the intercept is a positive number (notice the line crosses the y-axis above the x-axis, indicating a positive value of Y). We also know that it has a very high positive correlation because the points are very close to the line. Let's say it is represented by the following equation:

$$\mathrm{Y} = 2\mathrm{X} + 1.5$$

This suggests that as the x values increase by one unit, the y values increase by two units. Perhaps you recall a definition of the slope as $\frac{\text{rise}}{\text{run}}$; which is a shorthand way of saying that as the points rise a certain number of units along the y-axis, they also run a certain number of units along the x-axis.

Figure 2.1: Linear Regression

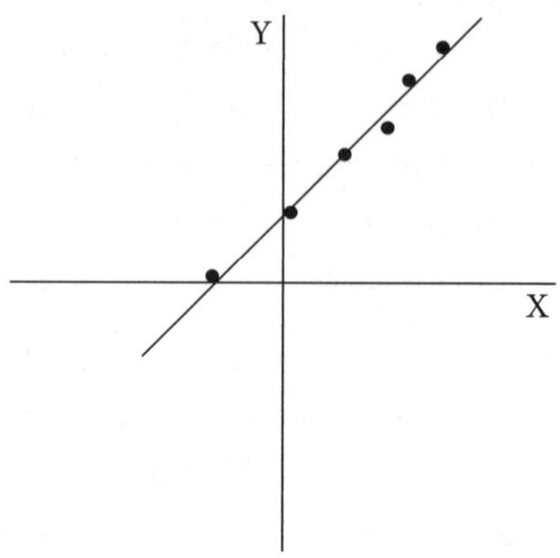

Unfortunately, real data never follow the kind of relationship we show in Figure 2.1 (all data points in a straight line). Nonetheless, in a simple linear regression analysis, we may represent the association between the variables by using a straight line. Whether a straight line is an accurate representation is a question we should always ask ourselves. Consider Figure 2.2. In SPSS, a similar graph may be computed with the General Social Survey (GSS) 2010 data by using *Graphs-Scatterplot-Simple* and placing hrsrelax on the y-axis and educ on the x-axis. In Stata, one could enter the following code:

```
twoway scatter hrsrelax educ || lfit hrsrelax educ
```

In Stata, the double bars (located on most keyboards directly below the Backspace key) tell the program to produce two graphs, one on top of the other. Here, we have asked for a scatter plot (scatter) and a linear fit line (lfit) in one graph. After entering the commands, a new window should open that displays a graph similar to that in Figure 2.2.

The association between the two variables shown in the scatter plot is positive, but notice that the points do not fall on the line that represents the slope. In fact, many points fall below the line and some are well above the line. The line represents a probabilistic statistical relationship, whereas the previous graph shows a mathematical relationship, which is deterministic. Therefore, we now must say that, *on average*, hours spent relaxing each day increase as educational attainment increases. The next question to ask is, "On average, how many units increase in education are associated with how many units increase in time spent relaxing?" This is a fundamental question that we try to answer with a linear regression model.

In a linear regression model, rather that $Y = mX + b$, we use the following form of the equation:

$$Y_i = \alpha + \beta_1 X_i$$

Figure 2.2: Linear Regression of Relaxation Time and Educational Attainment

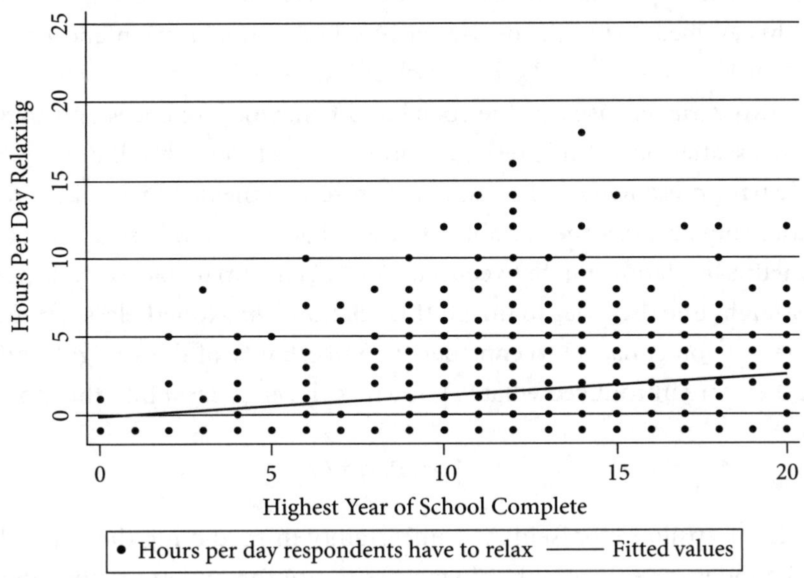

The Greek letter alpha represents the intercept (where the line crosses the y-axis) and the Greek letter beta represents the slope. These two terms are sometimes labeled *parameters*; however, in statistical analysis, parameters are usually defined as properties of a population, in contrast to properties of samples, which are labeled *statistics*. For the real stickler, perhaps we should not even use Greek letters in the equation because, technically speaking, they imply the use population parameters and not sample statistics. Hence, an alternative representation of the linear regression equation is:

$$y_i = a + b_1 x_i$$

Here, we used lowercase roman letters throughout the equation to represent a linear regression model that uses data from a sample. In reality, the equation using Greek letters and the equation using roman letters are used almost interchangeably. You will also notice the subscripted i; as before, this represents the fact that we have individual observations from the sample.

Another way of thinking about the slope, whether represented by the Greek letter beta or the letter b, is that it represents the change in y over the change in x. You may see this referred to as $\frac{\Delta y}{\Delta x}$, in which the Greek letter delta is used to denote change. This is just another way of saying rise over run. The intercept may also be defined as the predicted value of y when the value of x is zero.

You may have noticed a problem with the equations we have written so far. A hint is that we used the term *probabilistic* to describe the statistical relationship. If you are not sure what this all means, revisit the scatter plot between education and time relaxing mentioned earlier. Notice that the points clearly do not fall on the straight line. Take almost any two variables used in the social and behavioral sciences and they will fail to line up in a scatter plot. This presents a problem for those who desire exact models. Thus far, we have presented equations that require exactness, which misrepresents the actual relationship between the variables. Our straight line is, at best, an approximation of the actual linear relationship between x and y. Given a small data set, it is not difficult to draw a straight line that tries to match this relationship. Nonetheless, we must revise our linear regression equation to consider the uncertainty of the straight line. Statisticians have therefore introduced what is known as the *error term* into the model:

$$y_i = \alpha + \beta_1 x_i + \varepsilon_i$$

The Greek letter epsilon represents the uncertainty in predicting the dependent variable with the independent variable (if we were to use roman letters, epsilon would be replaced by e). Another way of thinking about the error term is that it represents how far away the individual y values are from the true mean value of Y for given values of x—something that is often referred to as a *residual.* Now, this last sentence is loaded with assumptions. We use, for instance, the term *true mean value* of Y. What does this mean, exactly? Well, think about our sample. If we have done a good job, our sample observations should represent some group from the population. For instance, if we have sampled from adults in the United States, some of our sample members should represent 25- to 30-year-olds. We assume that these sample members are a good representation of other 25- to 30-year-olds. Hence, their values on a characteristic such as education should provide a good estimate of mean education among those in this age group in the population. Suppose we wish to use parental education (x) to predict education (y) among this age group. Using the linear regression model, we assume that the error terms represent the distance from the points on the graph to the actual mean of y, or the population mean.

Unfortunately, we usually cannot test—at least not fully—the various assumptions we make when using the linear regression model. As we normally do not have access to the population parameters, we can only roughly test properties of the various assumptions we make when using linear regression. These assumptions include the following:

1. For any value of x, y is normally distributed and the variance of y is the same for all possible values of x. (Notice that we use uppercase letters to show that we are referring to the population.)

Figure 2.3: Visualizing Assumptions of Ordinary Least Squares Regression

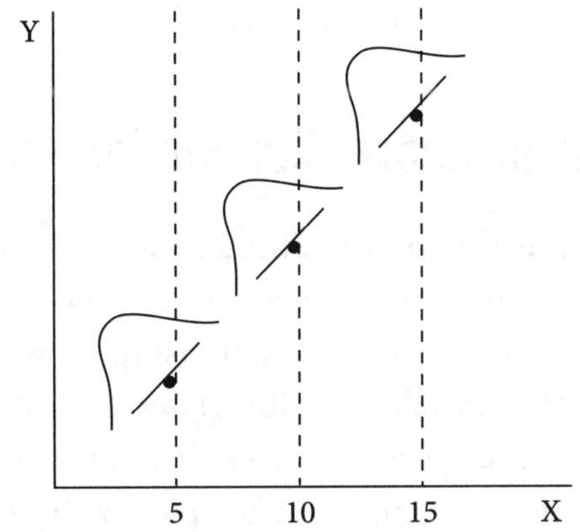

2. The mean value of Y is a straight line function of X. In other words, Y and X have a linear relationship.
3. The Y values are statistically independent of one another. Another way of saying this is that the observations are independent. For example, using the earlier example, we assume that our measures of public expenditures across the seven nations are independent. We usually assume that simple random sampling guarantees independence. (However, we should ask ourselves: Are the economic conditions of these nations likely to be independent?)
4. The error term is a random variable that has a mean equal to zero in the population and constant variance; this is implied by (1). Symbolically, this is shown as $\varepsilon \sim N(0,\sigma^2)$. As mentioned in chapter 1, the wavy line means "distributed as."

To test these assumptions, we should have lots of *x* values and lots of *y* values. For example, if we collect data on dozens of individuals with 10 years of education, then we expect the values of leisure time to be normally distributed within this value of the variable education. In other words, not everyone with 10 years of education will have the same amount of leisure time, but we would expect that it would be normally distributed within this particular value. Further, we expect the errors in predicting Y to have a mean of zero.

Figure 2.3 provides one way to visualize what is meant by some of these assumptions. We have sets of observations at 5, 10, and 15 years of education. We assume that the mean hours of leisure time for these three subsamples are a good representation of the actual means in the population. We also assume that leisure time at these three values of the independent variable are normally distributed, that the variances of each distribution are identical, and that the errors we make when we estimate the value of leisure time have a mean of zero (for example, the underestimation and overestimation that we make cancel out).

An Example of a Linear Regression Model Using SPSS and Stata

Perhaps you are confused at this point, although we hope not. Maybe an example using some data would be helpful. We turn again to our GSS 2010 data to help us. In this example, we are going to look at the relationship between days of poor mental health in the past 30 days (the variable mntlhlth) and educational attainment (educ). If you are using SPSS, you should first use the scatter plot drop-down menu (*Graph-Scatterplot-Simple*) to estimate a simple scatter plot with years of education as the independent variable (x) and days of poor mental health as the dependent variable (y). We do this because we think that education is a predictor of mental health and not vice versa. If you double-click the graph and highlight the data points, you can place a regression line in the graph using the icon labeled "add fit line" at the top of the window (we are assuming you are using SPSS 12.0 or higher). You should see a modestly decreasing slope, but notice how the points diverge from the line; only a few are relatively close to it.

Next, we will try a simple linear regression model using these variables. In SPSS, a linear regression model is set up using the *Analyze-Regression-Linear* drop-down menu. The window that opens up has the variables listed in a box on the left and several boxes on the right in which we will place the variables of interest. Notice that there is a box labeled *Dependent* and another labeled *Independent(s)*. We will ignore the other boxes, although we will learn more about them later. To estimate our model, place *mntlhlth* in the Dependent box and *educ* in the Independent(s) box. Then click *OK*. Your results should look like those in Table 2.1.

The analogous analysis in Stata begins by using the two-way plot command with the scatter and lfit plot types to estimate a simple scatter plot and linear fit line with poor mental health days (*mntlhlth*) as the outcome variable and years of educational attainment (*educ*) as the explanatory variable (`twoway scatter mntlhtlh educ || lfit mntlhtlh educ`). You should see a modestly decreasing slope, but notice how the points diverge from the line; only a few are relatively close to it.

Next, we will try a simple linear regression model using these variables. In Stata, a linear regression model is set up by using the regress command. The commands should look like this: `regress mntlhlth educ`. The output screen should then look like Figure 2.4.

Can we translate the numbers in this table into our linear regression equation? No matter if you used SPSS or Stata, we can, but first we need to find the slope and the

Table 2.1: SPSS Output

Model		Unstandardized Coefficients		Standardized Coefficients		
		B	Std. Error	Beta	*t*	Sig.
1						
	(Constant)	7.674	1.010		7.599	.000
	Highest year of school completed	−.275	.071	−.114	−3.894	.000

Note: Dependent variable = days of poor mental health past.

Figure 2.4: Stata Output

```
. reg mntlhlth educ
      Source |       SS           df       MS              Number of obs  =     1149
-------------+------------------------------------------   F(1, 1147)     =    15.17
       Model |  802.815106         1  802.815106           Prob > F       =   0.0001
    Residual |  60716.4295      1147  52.9349864           R-squared      =   0.0130
-------------+------------------------------------------   Adj R-squared  =   0.0122
       Total |  61519.2446      1148  53.5881921           Root MSE       =   7.2756

------------------------------------------------------------------------------
    mntlhlth |      Coef.   Std. Err.      t    P>|t|     [95 % Conf. Interval]
-------------+----------------------------------------------------------------
        educ | −.2747143   .0705415    −3.89   0.000    −.4131192   −.1363094
       _cons |   7.67381   1.009797     7.60   0.000     5.692553    9.655067
------------------------------------------------------------------------------
```

intercept. In SPSS, the term *coefficients* is used in a general sense, although the key columns are labeled *Unstandardized Coefficients* and *Standardized Coefficients*. We will focus for now on the former column as it provides the information we need. SPSS uses an uppercase B to indicate the intercept and slope, while Stata uses coef. It also uses the term *Constant* (in Stata, it is _cons) to label the intercept. Thus, our equation reads:

$$\text{Mental Health}(y_i) = 7.674 - \{.275 \times \text{Education } (x)\} + \varepsilon_i$$

The y and the x are included to remind you that poor mental health days is the dependent variable and education is the independent variable. Revisit the scatter plot between these two variables. The slope is negative. In other words, the line tilts downward when going from left to right. Our slope in the above equation is also negative but provides even more information about the presumed association between the variables.

Recall that the slope shows the average number of units increase in y for each one unit increase in x. This provides a way to interpret the slope in this equation:

Each one-year increase in educational attainment is associated with a 0.275-day decrease in number of days with poor mental health.

Of course, we need to specify how these variables are measured to determine what this statement means. For instance, because education is measured in years, we attach that unit of measure into the sentence, and because poor mental health is measured in days, we measure the dependent variable in that metric. It is important to attach a proper metric (dollars, days, points, years, and so forth) to the dependent and independent variables so that they make sense to an audience. Similarly, given our understanding of the intercept, we may say that the expected number of poor mental health days when education is zero is 7.674.

There are a few people who report that they have zero years of education, but it is exceedingly rare (only 0.25 percent of the GSS respondents said they had no education). Although zero years of education is possible, sometimes zero is an impossible number. For example, if we were focused on a sample of adults and had included age as our predictor, zero would be a meaningless number. In scenarios such as this, it is not a good idea to estimate values from your sample when no observations have this value. It can sometimes lead to ridiculous results if we try to make inferences too far outside our set of data values.

Keep in mind that we have used the term "associated" in our interpretations. Many people are uncomfortable with some of the terms used to describe the relationship between variables, especially when using regression models in the social and behavioral sciences. "Associated" is a safe term, one that implies that as one variable changes, the other one also tends to change. It should not be seen as a term denoting causation, though; each year increase in education does not lead to or cause an increase in the number of poor mental health days. We are sure you can think up several reasons why these two variables might be associated that have nothing to do with one variable causing the other.

In general, we say that the intercept represents the expected value of Y when X equals zero. The term *expected value* should remind you of the mean. The slope represents the expected difference or change in Y that is associated with each one-unit change or difference in X. We are inferring from the sample to the population here, although we do not have to. The two terms that are frequently used when considering the slope are *difference* and *change*. A problem with using the term *change* is that we do not necessarily observe changes; we (usually) note only differences in the values of variables.

Figure 2.5 shows a picture of the shifts that may be inferred from the linear regression model. With a little imagination, it also demonstrates another useful thing about this type of model: We may come up with predictions of the dependent variable. With a graph, we simply find that point along the x-axis that interests us, trace vertically to the regression

Figure 2.5: Visualizing the Regression Line

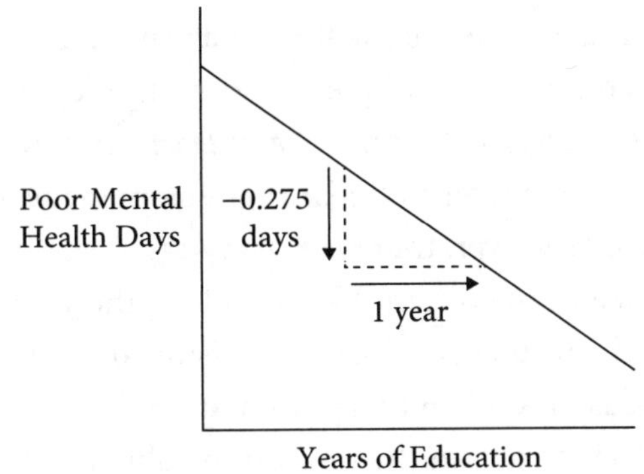

line, and then trace horizontally to the y-axis to find its value. This is known as the *predicted value* from the model. Graphs can be difficult to use to make precise predictions, however, so we may also use the regression equation to find predicted values. Suppose we wish to predict the number of poor mental health days for someone with 12 years of education:

$$\text{Mental Health}(y_i) = 7.674 - \{.275 \times 12\} + \varepsilon_i = 4.374$$

Hence, on average, individuals with 12 years of education have approximately 4.374 poor mental health days per month. We have used here the phrase "on average." This is another way of implying that for those individuals with 12 years of education, we estimate that the mean number of poor mental health days is 4.374. In other words, predicted values are *predicted means* of the dependent variable for particular values of the independent variable.

One of the questions you should ask yourself is why education is negatively associated with the number of poor mental health days. Actually, we should ask these types of questions before we estimate a regression model. Perhaps educated people have less to worry about, are less stressed, and have better mental health as a result. As suggested earlier, could there be a third variable—such as gender, race, personality, or some other factor—that explains the association between the two variables? At this point, we cannot tell, but we should consistently seek reasonable answers to why two or more variables are associated, and we should be willing to revise our conclusions as new evidence emerges.

Formulas for Slopes and Intercepts

You should be asking at this point: From where do these slopes and intercepts come? Certainly, we can plot the data—by hand if the sample is small enough—and then guess which straight line best represents the association between two variables. It makes sense to try to draw a line that is as close as possible to all the points. Another way of thinking about this is to try to draw a line that, if we calculated an aggregate measure of distance from the points to the line, will have as small a value as possible. This is precisely what the most common technique for fitting the regression line does. Known as *ordinary least squares* (OLS) or the *principle of least squares*, this technique is used so frequently that many people refer to linear regression analysis as OLS analysis or as an OLS regression model. However, there are many types of estimation routines, such as weighted least squares and maximum likelihood, so there are actually a number of ways we might compute the results of a linear regression analysis. Because it is so widely used, though, we will focus on OLS in this presentation.

The main goal of OLS is to obtain the minimum value for the following equation:

$$\text{SSE} = \sum (y_i - \hat{y}_i)^2 = \sum (y_i - \{\alpha + \beta_1 x_i\})^2$$

The quantity SSE is the *Sum of Squared Errors* or the *Residual Sum of Squares*. There is also a new symbol that we have not seen before: $\hat{y}$ (also called y-hat). This is the symbol for predicted values of y based on the model that uses the sample data. The goal of OLS is to minimize the SSE, or make it as close to zero as possible. In fact, if the SSE equals zero, then the straight line fits the data points perfectly; in other words, they fall right on the line. In addition, if the points fall directly on the line, the Pearson's correlation coefficient is 1 or −1 (depending on whether the association is positive or negative). Another new term is *residual*: It is a common name for a measure of the distance from the regression line to the actual points, so there are actually many residual values that result from the model. It is typically represented by *e* to indicate that it is another way of showing the errors in estimation $\{residual_i(\varepsilon_i) = (y_i - \hat{y}_i)\}$. As we will learn in later chapters, the residuals are used for many purposes in linear regression models.

As mentioned earlier, the straight regression line almost never goes through each data point. Rather, there are discrepancies from the line for two reasons: (1) There is simply variation of various types whenever we measure something; and (2) a straight line association is not appropriate. The former situation is a normal part of just about any linear regression analysis. Although we wish to minimize errors, there is usually natural or random variation because of the intricacies and vicissitudes of behaviors, attitudes, and demographic phenomena. The second situation is more serious; if a straight line is not appropriate, we should look for a relationship—referred to as *nonlinear*—that is appropriate. Chapter 10 provides a discussion of what to do when we find a nonlinear relationship among variables.

But how do we come up with an equation that will minimize the SSE and thus allow us to both fit a straight line and estimate the slope and intercept? Those of you who have taken a calculus course may suspect that, because we are concerned with changes in variables, derivatives might provide the answer. You are correct, although through Cramer's rule and the methods of calculus some simple formulas have been derived. Among the various alternatives, the following least squares equation for the slope has been shown to be the best at minimizing the SSE:

$$\hat{\beta} = \frac{\sum (x_i - \bar{x})(y_i - \bar{y})}{\sum (x_i - \bar{x})^2}$$

This equation works very well if the assumptions discussed earlier are met. We will have a lot more to say about assumptions in later chapters. The numerator and denominator in this equation should look familiar. The numerator is part of the equation for the covariance (see chapter 1), and the denominator is the formula for the sum of squares of x. Here, we use the common x-bar and y-bar symbols for the means, and we use $\hat{\beta}_1$ to indicate that we are estimating the slope with the equation. Let's think for a moment about what happens as the quantities in the equation change. Suppose that the covariance increases. What happens to the slope if the sum of squares does not also change? Clearly, it increases. What happens as the sum of squares of x increases? The slope decreases. So, all else being equal, as the variation of the independent variable increases, the slope decreases. Typically, we wish to have a large positive or negative slope if we hypothesize that there is an association between two variables.

Now that we have the slope, how do we estimate the intercept? The formula for the intercept is the following:

$$\hat{\alpha} = \bar{y} - \{\hat{\beta}_1 \times \bar{x}\}$$

The estimated intercept is computed by using the means of the two variables and the estimated slope. We should note at this point that we rarely have to use these equations because programs such as SPSS will conduct the analysis for us. Moreover, programs such as SPSS and Stata do not use these equations directly; rather, they use matrix routines to speed up the process. It is useful, though, to practice using these equations with a small set of data points.

Hypothesis Tests for the Slope

Now that we have seen examples of slopes and intercepts, as well as where they come from, we should consider the issue of hypothesis testing. What are the most common hypotheses we wish to examine using a linear regression model? Generally speaking, the most common is that there is a linear association between the independent and the dependent variables. But we should make this more precise. Using your imagination, common sense, understanding of the research on your topic, and perhaps even colleagues' ideas as you discuss your research plans, you should try to specify the reason there might be an association and then write down the null and alternative hypotheses. Let's assume that all of these things tell us that education should be positively associated with the number of poor mental health days. This means that we think there is a positive slope in a regression model that considers these variables. Our hypotheses are therefore,

$$\text{H}_0 : \text{linear regression slope} \leq 0$$
$$\text{H}_a : \text{linear regression slope} > 0$$

Although these are reasonable hypotheses, for some reason, most researchers who use linear regression models define the hypotheses more generally as saying either that the slope is zero or that the slope is not zero. This, of course, means that the alternative hypothesis is that the slope is simply different from zero and that the null hypothesis is that the slope is zero: $\{\text{H}_a : (\beta_1 \neq 0)\}$ versus $\{\text{H}_a : (\beta_1 = 0)\}$.

It may appear obvious that we can simply look at the slope coefficient and determine whether it is zero, but let's not forget a crucial issue. Remember that we are assuming that we have a sample that represents some target population. Ours is only one sample among many possible samples that might be drawn from the population. Perhaps we were just lucky (or unlucky, depending on how you look at things) and found the one sample among countless others that had a positive slope. How can we be confident that our slope does not fall prey to such an event? Although we can never be absolutely certain, we may use a significance test or CIs to estimate whether the slope in the population is likely to be zero.

It is now time to return to standard errors. In chapter 1, we learned how to compute the standard error of the mean and how this statistic is interpreted. There is also a statistic known as the *standard error of the slope* that is interpreted in a similar way. That is, the standard error of the slope estimates the variability of the estimated slopes that might be computed if we were to draw many, many samples. Let's say we have a population in which the association between age and alcohol consumption is zero. That is, the correlation is zero and the population-based linear regression slope is zero. Therefore, the null hypothesis of a zero slope is true ($\text{H}_0 : \beta = 0$). If we were to draw many samples, can we infer what percentage of the slopes from these samples would fall a certain distance from the true mean slope of zero? It turns out we can because the linear regression slopes from samples, if many samples are drawn randomly, follow a t distribution. We therefore know, for instance, that in a sample of, say, 1,000, we would expect only about 5 percent of the slopes to fall more than 1.96 t values (similar to z values in the standard normal distribution) from the true mean of zero. We should only occasionally find a slope this far from zero if the null hypothesis is true. If the null hypothesis is not true (or if we may *reject it*, to use a common phrase), then a slope this many *t* values from zero should be relatively common. The question then becomes, how can we compute these *t* values?

The *t* value in a linear regression model is computed by dividing the estimated slope by the estimated standard error. We have already seen how to compute a slope in a simple linear regression model. Here is the formula for the standard error:

$$SE(\hat{\beta}_1) = \sqrt{\frac{\sum (y_i - \hat{y}_i)^2 / n-2}{\sum (x_i - \overline{x})^2}} = \sqrt{\frac{\text{SSE} / n-2}{SS[x]}}$$

This equation provides the standard error formula for the slope in a simple linear regression model only. As with the slope equation, the standard error equation includes some familiar elements. In the numerator, for instance, we see the sum of squared errors (SSE) and the sample size (*n*); in the denominator, we find the sum of squares of *x*. As the sum of squares of *x* gets larger, the standard error becomes relatively smaller. As the SSE becomes larger, the standard error also becomes larger. This should not be surprising if we think about our scatter plot: Larger SSEs indicate more variation about the regression line. Therefore, our uncertainty about whether we have a good prediction of the population slope should also increase. However, what happens to the standard error as the sample size increases? Working through the algebra implied by the equation, we can see that the standard error decreases with larger samples. Again, this should make sense: The larger the sample, the more it is like the population from which it was drawn, and the more confidence we should have about the sample slope. In fact, some observers complain because if we have a larger enough sample, even one that was not drawn randomly, we can make claims of certainty that are not justifiable.

As mentioned in the beginning of the last paragraph, a *t* value in a linear regression model is computed by dividing the slope by the standard error:

$$t \text{ value} = \frac{\hat{\beta}_1}{se(\beta_1)}$$

Occasionally, you might see one beta value minus another in the *t*-value equation (for example, $\hat{\beta}_1 - \beta_0$). Although it may be used in such a manner, it is much more common to assume that we wish to compare the slope from the regression model to the slope implied by the null hypothesis, which is typically that the slope is zero in the population. Each *t* value is associated with a *p* value that depends on the sample size. This is the basis for using significance tests to determine the importance of the slope. Fortunately, we do not need to use *t* tables to determine the *p* value. Statistical software provides the associated *p* values.

Recall that we used SPSS and Stata to estimate a model with poor mental health days as the dependent variable and years of education as the independent variable (these results are represented in Table 2.1 and Figure 2.4, respectively).

We have already learned to interpret the slope and intercept in this model. But notice that both programs also compute the standard errors (listed as Std. Error in SPSS and Std. Err. in Stata) of the slope and intercept, the *t* values (under the column labeled *t*), and the *p* values (under the column labeled Sig., an abbreviation for *significance*). The *t* value for the slope is $-.275/.071 \approx -3.9$. A *t* value of this magnitude from a sample of 1,149 observations has a *p* value of less than .001. Although it did not happen here, sometimes the two programs will produce slightly different results for the same regression—this is because they use slightly different algorithms to come up with the results. However, they are substantively similar to each other. Although it is rarely used, there is also a *p* value for the intercept. It compares the intercept to the null hypothesis value of zero.

We usually do not have need to interpret the standard error of the slope or intercept directly, nor do we interpret the *t* value. Rather, we may focus on the *p* value. Unfortunately, *p* values are frequently misused. You might recall that most statistical analyses use a threshold or rule of thumb that states that a *p* value of .05 or less means that the statistic is somehow important. Actually, this is an arbitrary value that has, for various reasons, become widely accepted. In fact, when one finds a *p* value of .05 or less, it is common to claim that the slope (or other statistic) is *statistically significant*, but what precisely does the *p* value mean? Can we come up with an interpretation? The answer is yes, but we must again assume that the sample slope says something reasonable about the true population slope. Suppose, for instance, that we find a *p* value of .03. The following interpretation is reasonable (even if it is not entirely persuasive): If the actual slope in the population is zero, we would expect to find a sample slope this far or farther from zero only three times out of every 100 samples, if we were to draw many samples from the population.

You should notice that this interpretation construes the *p* value as a probability. Another way of thinking about this *p* value is that it suggests (but, of course, does not prove) that, if the null hypothesis is true in the population, we would reject it only about 3 percent of the time given a slope this far or farther from zero. Once again, we are assuming that we could draw many samples to reach this conclusion.

In the model of mental health and education, the *p* value suggests that if the population slope is zero, we would expect to find a slope of $-.275$ or some value farther from zero less than one time out of every 1,000 samples drawn. This is clearly a small *p* value and offers a large degree of confidence that there is a nonzero statistical association between mental health and education.

However, note that we use the phrase "something farther from zero." This implies that we are using what we labeled in chapter 1 as a *two-tailed significance test*. Imagine a bell-shaped curve with the value of zero as its middle point. We may then picture the area under the curve as representing the frequency of slopes from a large number of samples. The mean of zero represents the null hypothesis. Hence, this bell-shaped curve signifies the distribution of slopes for the null hypothesis. A certain percentage of the total area of the curve lies in the tails of this distribution. For instance, if we have a large sample (say, 200 or more), then we know that 2.5 percent of the total area falls in each tail marked by the values ±1.96. A two-tailed *p* value of .05 implies a *t* value of 1.96 (for large samples). When we rely on programs such as SPSS and Stata for *p* values, they typically provide two-tailed tests: They assume that the slope can be in either of the tail areas, above or below the mean. In other words, a negative slope and a positive slope are equally valid for rejecting the null hypothesis as long as their *t* values are sufficiently large (in absolute value).

However, suppose our alternative hypothesis indicates direction, such as that the slope is greater than zero. We are then justified in using a *one-tailed test*: We are concerned only with the conceivable slopes that fall in the upper tail of the *t* distribution. The 5 percent area that most people use falls above the mean of zero; in large samples, this has a threshold at 1.64 *t* units above the mean. How can we use a one-tailed test when programs such as SPSS or Stata assume we want two-tailed tests? We may take the *p* value provided by either program and divide it in half to reach a conclusion about the statistical significance of the slope. This can quickly become confusing, especially as we think about taking only one sample and then trying to infer something about a population. Perhaps it is easiest to simply remember that we wish to have slopes far from zero if we want to conclude that there is a nonzero linear association between an independent and a dependent variable.

As with other statistical tests that compare a value computed from the data to a hypothetical value from a null hypothesis, there are many researchers who argue that significance tests that use only *p* values are misleading: They deceive us into thinking that our estimates are precise and fail to account for the uncertainty that is part of any statistical model. Many recommend that we should, therefore, construct CIs for slopes, much like they are constructed for means. Because we have point estimates and standard errors, constructing CIs for linear regression slopes is relatively simple. In fact, the same general formula we used in chapter 1 is applicable:

$$\text{Point estimate} \pm [(\text{confidence level}) \times (\text{standard error})]$$

As discussed in chapter 1, the confidence level represents the percentage of the time we wish to be able to say that the interval includes the point estimate. Given the special nature that the .05 *p* value has taken on, it is not surprising that a 95 percent CI is the most common choice. If we wish to use a two-tailed test (again, the most common choice), then, assuming a large sample, the value of 1.96 is used for the confidence level. For smaller samples, you should consult a table of *t* values or ask the statistical software to compute CIs for the slopes. Here is an example from our linear regression model:

$$-.275 \pm [1.96 \times 0.071] = \{-.413, -.136\}$$

In this example, we used a confidence level of 1.96. This is found in a table of *t* values (the *df* is $n - 2$) and is associated with 2.5 percent of the area in each tail of the distribution. In other words, our confidence level is 95 percent. The interpretation of this CI is similar to the interpretation offered in chapter 1: Given a sample slope of 14.459, we are 95 percent confident that the population slope representing the association between education and poor mental health days falls in the interval bounded by −.413 and −.136. (Note: SPSS will compute CIs for slopes if you check the appropriate box in the Statistics window. Stata automatically produces the CI. The resulting intervals are slightly different than those we just computed because of rounding error.) If we wish to obtain a higher degree of confidence, we could increase the CI. For example, we could use a 99 percent CI. This would make us 99 percent confident about where the population slope falls—though the interval would be wider because the *t* value used in the equation is larger.

An interesting relationship between *p* values and CIs occurs when using the same *t* value for determining the threshold and for the confidence level. Suppose we rely on 1.96 to compute a *p* value of .05 (two-tailed test) and for our confidence level. Then, if the slope divided by the standard error—that is, the *t* value from our regression model—is less than 1.96, the *p* value is greater than .05. In this situation, the CIs that use 1.96 to represent the confidence level include zero. Thus, we would conclude that the null hypothesis cannot be rejected. In other words, our conclusions about statistical significance are generally the same whether we use *p* values or CIs. The main advantage of CIs is that they are a better reflection of the uncertainty that is a fundamental part of statistical modeling.

Key Terms

The following is a list of key terms and concepts in chapter 2. You can find definitions to these words throughout the chapter.

- Bivariate regression
- Coefficients
- Deterministic
- Error term
- Expected value
- Intercept
- Nonlinear
- Ordinary least squares (OLS)

- *p* value
- Parameters
- Predicted means
- Probabilistic
- Residual
- Slope
- Standard error of the slope
- Statistics
- Statistically significant
- Sum of squared errors
- *y*-hat

Analysis Exercise

1. Using either SPSS or Stata, run a bivariate regression with days of poor physical health (physhlth) as the dependent variable and days of poor mental health (mntlhlth) as the independent variable. Then do the following:
   a. Interpret the coefficient for mental health.
   b. Consider whether the effect of mental health on income is statistically significant. At what level?
   c. Interpret the intercept of the model.
   d. Write out the correct equation for this model.

Solution

a. A one-day increase in days of poor mental health is associated with a 0.328-day increase in poor physical health days.
b. Yes, the effect is statistically significant. The *p* value is .000, which means that the relationship is significant at $p < .001$.
c. The intercept of the model is 0.709, meaning that when days of poor mental health is zero, the expected number of days with poor physical health is 0.709.
d. The correct equation is $Y = 0.709 + 0.328(\text{mntlhlth}) + e$.

CHAPTER 3 MULTIPLE LINEAR REGRESSION ANALYSIS

Chapter Objectives

This chapter covers multiple regression analysis, which includes several independent variables as predictors of the dependent variable. It provides an overview of the model that is the backbone of statistical analysis in social work and the social sciences.

In chapter 3, we will

- introduce you to the multiple regression model, which includes several independent variables to predict an outcome
- discuss the concept of statistical control
- briefly introduce you to the assumptions of linear regression
- discuss why regression is a good statistical method.

We have now learned about some of the basic aspects of linear regression analysis, in particular where the slopes and intercepts come from, how to interpret them, and how to determine whether they are meaningful. However, we considered only the simplest case, that of a model with only one explanatory variable. We are now prepared to consider linear regression models that include more than one explanatory variable. We will continue to consider many of the same issues as with the simpler model, but as we shall learn, a key difference between simple linear regression models and multiple linear regression models is in the interpretation of the coefficients.

A simple linear regression model is usually not very interesting. For instance, we saw in the last chapter that there is a negative association between poor mental health days and education. However, you do not need to be a social worker or social scientist to realize that there might be other factors that are related to poor psychological health. Some of these factors include race, gender, parenthood, difficult life situations, and genetic predisposition; we could go on and on. The main point is that there are usually many potential explanatory variables used to predict an outcome variable. Theories of various stripes should be used to guide the selection of variables; yet it is rare that these theories can be narrowed down to only one explanatory variable.

A key reason to include other variables in the model is because they may account for the association between one of the explanatory variables and the outcome variable. This issue is known as *confounding*, because we say that if variable x_2 accounts for the association between x_1 and y, then x_2 is said to *confound* the association. For example, suppose we are interested in the association between the frequency of purchasing lighters and the rate of lung disease across a sample of U.S. counties. We would likely find a positive association between these two variables. Would we, therefore, conclude that purchasing lighters leads to lung disease? Probably not, especially when we realize that cigarette smoking is related to both purchasing lighters and lung disease. We say that cigarette smoking is a confounding variable and that the association between lighters purchased and lung disease is *spurious* (see chapter 7). Hence, smoking should be included in a regression model that predicts lung disease, especially if the model also includes the frequency of lighter purchases. (Actually, it should always be used in models predicting lung disease.)

An Example of a Multiple Linear Regression Model

Understanding the complex way that multiple linear regression models estimate slopes and other statistics can be difficult. To estimate this model, we continue to use ordinary least squares (OLS), but the formulas become more complex as we add explanatory variables. It is perhaps best for now to see one of these models and then discuss how to use it. We will go back to the GSS 2010 data file once more but consider some other variables. Specifically, the data set has a measure of days of poor physical health in the last 30 days (*physhlth*), income (*realrinc*), and age (*age*). Before estimating a multiple linear regression model, it is useful to check the correlations among the variables to see, in a rough sense, the strength of their associations. SPSS returns the output shown in the table on the next page when we use the following drop-down menu: *Analyze-Correlate-Bivariate*; and place the three variables in the *Variables* box. Stata returns the output shown in Figure 3.1 when the following command is used: `pwcorr mntlhlth educ physhlth realrinc age, sig.` (Note: `pwcorr` asks for pairwise correlations, and `sig` requests p values.)

For now, we are going to focus on the results produced by Stata (shown in Figure 3.1), although SPSS produces an almost identical table. Note that there are some interesting and statistically significant correlations among the variables. The number of poor physical health days is positively associated with poor mental health ($r = .3731$; $p < .001$), whereas age is negatively correlated ($r = -.1658$; $p < .001$). Other significant correlations exist as well: Income and education are positively correlated, whereas age is negatively correlated with physical health and positively correlated with income. There is also a statistically significant correlation between mental health and education

Figure 3.1: Bivariate Correlations Produced by Stata

	mntlhlth	educ	physhlth	realrinc	age
mntlhlth	1.0000				
educ	−0.1142	1.0000			
	0.0001				
physhlth	0.2644	−0.0764	1.0000		
	0.0000	0.0094			
realrinc	−0.1078	0.3911	−0.0853	1.0000	
	0.0008	0.0000	0.0079		
age	−0.0978	−0.0490	0.0214	0.1844	1.0000
	0.0009	0.0271	0.4674	0.0000	

($r = -.088$; $p = .003$). Thus, if our hypotheses propose that education, physical health, income, and age are explanatory variables that predict mental health, we have some preliminary evidence that the key variables are associated.

Let's now see what a linear regression analysis tells us about these associations. We will begin with a simple linear regression model using only age as an explanatory variable. By now, you should be somewhat familiar with Stata and SPSS output, so we are going to simply show the results in Table 3.1.

How do we interpret the slope associated with age? It should be a little more familiar now. We may say, "Each one-year increase in age is associated with a decrease of 0.052 poor mental health days." As we noted earlier, there can be slight differences between Stata and SPSS estimates, which we notice here. The *p* value suggests that, if the actual population slope is zero, we would expect to get a slope of 0.052 or something farther away from zero less than one time out of every 1,000 samples. The intercept indicates that the expected number of poor mental health days is around six when the age is zero (is this a reasonable number?).

If you are using SPSS, you may have noticed that there is one entry in the table that we have not yet discussed. Under the column labeled Standardized Coefficients, SPSS lists the term "Beta" and, below it, the number −.094. At this point, we should discuss standardized coefficients. To calculate standardized coefficients using Stata, use the regress command with the beta option (`regress mntlhlth age, beta`). Under the column labeled beta, Stata lists the number −.098. Have you seen this number before? Look at the correlation matrix in Figure 3.1. The Pearson's correlation between age and poor mental health days is −.098. The standardized coefficient measures the association between the variables in term of standard deviation units or *z* scores; when there is only one explanatory variable, this statistic is the same as the Pearson's correlation

Table 3.1: Days of Poor Mental Health Regressed on Age

Variable	*b*	*SE*
Age	−0.052	0.015**
N	1,150	
R^2	.01	

$^{**}p < .01$.

(recall from chapter 1 that we identified the correlation as the standardized version of the covariance). As we shall soon see, this will change once we move to a multiple linear regression model.

Standardized regression coefficients—or *beta weights* as they are often called—may be interpreted using standard deviations. One way to consider them is to imagine transforming both the explanatory variable and the outcome variable into *z* scores (that is, standardize them), running the regression model, and looking at the slope. This slope represents the association between variables in standard deviation units and is identical to the standardized regression coefficient that Stata produces. Hence, our interpretation of the standardized regression coefficient in the violent crimes model is "Each one standard deviation increase in age is associated with a .098 standard deviation decrease in the number of poor mental health days."

The standardized regression coefficient and the unstandardized regression coefficient are related based on the following formula:

$$\beta^* = \beta\left(\frac{s_x}{s_y}\right) \quad \text{where } \beta^* \text{ is the standardized coefficient}$$

Recall that s_x denotes the standard deviation of x and that s_y denotes the standard deviation of y. In the earlier example, we may use this formula to compute the standardized coefficient for age:

$$-0.098 = -.0517 \times \left(\frac{43.721}{3.825}\right)$$

There are some researchers who prefer standardized coefficients because they argue that these statistics allow comparisons of explanatory variables within the same linear regression model. An explanatory variable with a standardized coefficient farther away from zero, they argue, is more strongly associated with the outcome variable than are explanatory variables with standardized coefficients closer to zero. However, this assumes that the distributions of the explanatory variables are the same. Many times,

Table 3.2: Days of Poor Mental Health Regressed on Age and Education

Variable	*b*	*SE*
Age	−0.052	0.015**
Education	−0.264	0.070***
N	1,148	
R^2	.02	

**$p < .01$. ***$p < .001$.

though, one variable is much more skewed than another, so a standard deviation shift in the two variables is not comparable. When we learn about dummy variables in chapter 6, we will also see that standardized regression coefficients are of limited utility. Moreover, standardized regression coefficients should not be used to compare coefficients for the same variable across two different linear regression models.

We will now estimate a multiple linear regression model by adding education to the previous model. In SPSS, we simply add this new variable in the independent variables box, whereas in Stata, we simply add the new variable to the command (`regress mntlhlth age educ`). Table 3.2 reports the results of this regression. There are now two slopes, one for each explanatory variable. We will first offer some interpretations for the numbers in this table and then try to understand how a multiple linear regression model comes up with them.

Try to imagine the association among these three variables in three dimensions. If this is difficult, begin with a two-variable scatter plot with *x*- and *y*-axes and then picture another axis, labeled *z*, coming out of the page. The three axes are at 90° angles to one another. Now, how is the intercept interpreted in this representation? Perhaps you can guess that it is the expected value of the outcome variable when both of the explanatory variables are zero. If we had an individual who was just born and had no schooling, we would expect him or her to have 9.555 poor mental health days in a month. Of course, the intercept in this model is meaningless as it falls so far outside of any reasonable values of the explanatory variables.

The slopes are interpreted the same way as before, but we add a phrase to each statement. Here is an example: "Controlling for age, a one-year increase in education is associated with a 0.264-day decrease in poor mental health." Notice that we have used the phrase "controlling for." Other phrases that are often used include "statistically adjusting for" or "partialling out." Because we are statistically controlling for or partialling out the effects of a third variable, multiple linear regression coefficients are also known as *partial regression coefficients* or *partial slopes*.

Figure 3.2: Visualizing Statistical Control

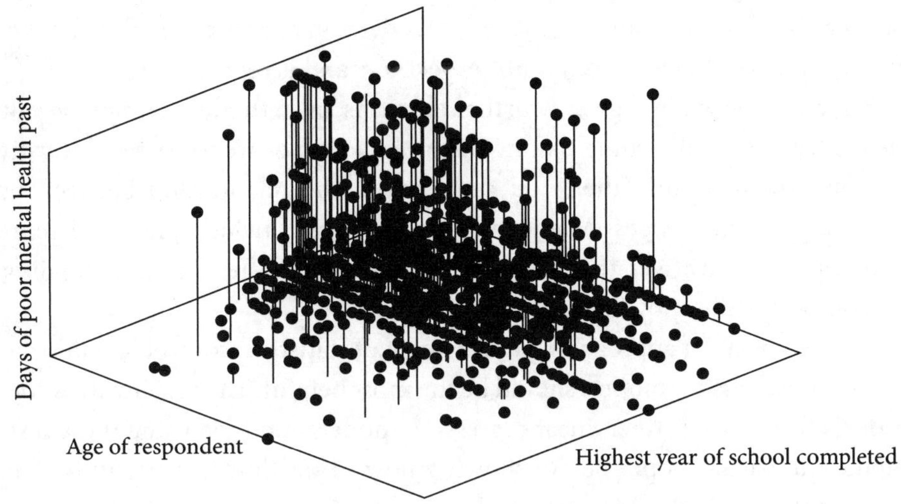

Statistical control or adjustment is a difficult concept to grasp. One way to think about it is to claim that we are estimating the slope of one variable on another while holding constant the other explanatory variable. For instance, when we claim that each year increase in education is associated with 0.264 fewer poor mental health days, we are assuming that this average increase occurs for any age, whether it is 18 or 80. In other words, age is assumed to be held constant. To further our understanding of this important idea, we will discuss four ways of approaching this topic, with the hope that perhaps one or two of these will be useful. However, the first way may only be useful if you have superior spatial perception skills. This is to use a three-dimensional (3-D) plot to visualize the relationship among the three variables. SPSS has a 3-D graphing option. By plotting three variables and then using the regression option to calculate the slopes, you may be able to visualize the meaning of statistical control. SPSS's 3-D option allows you to rotate the graph, so it can be seen from different angles. Looking at how the points follow a particular variable's slope may allow you to visualize the meaning of statistical control. Stata has a user-written 3-D graphing option (called `scat3`; type `find it scat3` on the command line). By plotting three variables (type `scat3 educ age mntlhlth` for this example) and then using the regression option to calculate the slopes, you may be able to visualize the meaning of statistical control. Looking at how the points follow a particular variable's slope may allow you to visualize the meaning of statistical control. This graph is in Figure 3.2.

You may find that this graph is not particularly helpful. A second way to understand statistical control assumes experience with calculus. A partial slope is synonymous with a partial derivative. Suppose that y is a function of two variables, x and w, that we wish to represent in a regression equation. If w is held constant (for example, $w = w_0$), then y becomes a function of a single variable x. Its derivative at a particular value of x is known as the *partial derivative* of y with respect to x and is represented by $\frac{\partial y}{\partial x}$ or $\frac{\partial f(x,w)}{\partial x}$ where $y = f(x, w)$. We will not go any further than this, even though it might be valuable to some readers. We will assume that if you have had exposure to partial derivatives—and you find them useful—then you can explore the relationship between partial derivatives and partial slopes. Many calculus textbooks include graphical representations of partial derivatives that may help you visualize the notion of holding one variable constant while allowing another to vary.

The next two ways of approaching the concept of statistical control or adjustment in multiple linear regression models should be the most helpful. The first involves computing residuals from two distinct linear regression models and then using these residuals to compute the partial slope. Before showing how to do this, it is useful to consider the nature of the residuals. As mentioned in chapter 2, residuals measure the distance from the regression line to the actual observations. A basic formula for a residual is $\text{resid}_i = (y_i - \hat{y}_i)$. (As mentioned earlier, you will often find the residuals represented as e_i.) This means that we take the observed values and subtract the values we predict based on the linear regression model. As indicated by the subscripts, we do this for each observation. Another way to think about residuals is that they measure the variation that is left over in the outcome variable after we have accounted for the systematic part that is associated with the explanatory variable. Part of what is left over may be accounted for by another explanatory variable. This part is represented by the partial slope.

To understand how this is done in SPSS, take the following steps:

1. Estimate a simple linear regression model with mental health as the outcome variable and education as the explanatory variable (leave age out of the model). After placing these variables in the dependent and independent variable boxes, click the lower right-hand box that is labeled *Save* and then, in the window that opens, check the box near the top that is labeled Residuals–Unstandardized. This will compute a new variable that is made up of the residuals from the model for each observation. These residuals represent the variability in poor mental health days that is not accounted for by education.
2. Estimate a second regression model with age as the outcome variable and education as the explanatory variable. Once again, save the residuals from this model. These residuals represent the variability in the age that is not accounted for by education.
3. Estimate a third regression model that uses the residuals from the first step as the dependent variable and the residuals from the second step as the independent variable. There is no need to save the residuals from this model.

Figure 3.3: Visualizing the Slope Coefficient

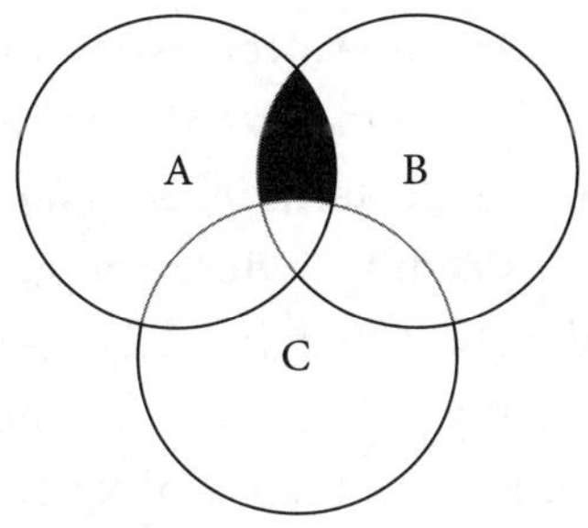

To understand how this is done in Stata, take the following steps:

1. Estimate a simple linear regression model with mental health as the outcome variable and education as the explanatory variable (leave age out of the model). Next, use the *predict* command to calculate the unstandardized residuals (predict resid, residual). This will compute a new variable (called *resid* in our example) that is made up of the residuals from the model for each observation. These residuals represent the variability in mental health days that is not accounted for by education. (Browse Stata's Data Editor to examine these residuals.)
2. Estimate a second regression model with age as the outcome variable and education as the explanatory variable. Once again, predict the residuals from this model (use a different variable name, such as *resid1*). These residuals represent the variability in age that is not accounted for by education.
3. Estimate a third regression model that uses the residuals from the first step as the outcome variable and the residuals from the second step as the explanatory variable (for example, regress *resid* and *resid1*). There is no need to predict the residuals from this model.

Now, look at the unstandardized regression coefficient from the third model. It is −.049, the same number we found when we estimated the multiple linear regression model earlier. This is because the slopes from these two models represent the same thing: the covariability between age and poor mental health that is explanatory of education.

The fourth way to consider statistical control is similar to using residuals, except it is visual. Figure 3.3 shows three overlapping circles labeled A, B, and C. We could have labeled them y, x_1, and x_2 to show that they are variables we are interested in using in a regression model. Let's assume that A represents the outcome variable and that B and C are explanatory variables. The total area of the circles represents their variability (if you find it more comforting, think of the sums of squares or the variances as represented by circles). The overlapping areas signify joint variability, such as a measure of covariance between two variables. Notice the darkened area of overlap between A and B. It does not include any part of C. This area denotes the joint variability of A and B that is not accounted for at all by the explanatory variable C. It represents the partial slope or the regression coefficient from a multiple linear regression model.

Notice that there is no need for B and C to be completely independent. Rather, we can assume that they covary. Hence, the term *independent variable*, which is often used in regression modeling, can be misleading: It does not mean that the predictor variables are independent from one another, only that they might—assuming that our conceptual model is a good one—independently predict the outcome variable. Extending multiple linear regression models to include more than two explanatory variables is easy and requires no deeper level of understanding than that which we already possess. The phrase "statistically adjusting for the effects of" must now either specify each other explanatory variable or state something such as "statistically adjusting for the effects of the other variables in the model, each one-unit increase in variable x_1 is associated with a [partial slope] unit increase in variable y."

A Brief Overview of the Assumptions of Multiple Linear Regression Models

Now that we have some sense of what multiple linear regression models produce in the way of slopes, intercepts, and p values, it is a good time to review the assumptions that we make when using the model. We saw some of these assumptions in chapter 2, but in the interest of completeness, we will revisit them. This is a brief overview, however, because several of the subsequent chapters review these assumptions in much more detail.

1. *Independence*: The y values are statistically independent of one another. We saw that this assumption applies to the simple linear regression model. Now that we have seen a couple of regression models, we can think about how this assumption might operate. For example, when analyzing state-level data, we assume that our measures are independent across states—but is this true? It is likely that states that share borders are more similar in a lot of ways than states that are in different parts of the country.
2. *Distribution of the error term*: The error term is normally distributed with a mean of zero and constant variance. The first part of this assumption—that the mean is zero—is important for estimating the correct intercept, but this assumption taken as a whole is also important for making inferences from the sample to the population. In other words, it is important for significance tests and confidence intervals.
3. *Specification error*: The correlation between each explanatory variable and the error term—or residuals—is zero. If there is a noticeable correlation, we have left something important out of the model. In chapter 7, we will learn that

this problem involves whether we have *specified* the correct model; hence, it is known as *specification error*.

4. *Measurement error*: The *x*s and *y* are measured without error. Notice that we use the lowercase to indicate the explanatory and outcome variables. This is because we are concerned primarily with sample measures. As the name implies, we must consider whether we are measuring the variables accurately. When we ask people to record their family incomes, are they providing us with accurate information? When we ask people whether they are happy, might some interpret this question differently than others? We will learn in chapter 8 that measurement errors are a plague on the social and behavioral sciences.
5. *Collinearity*: There is not perfect collinearity among any combination of the *x*s. If you consider the overlapping circles that we saw in the last section, you can visualize what this means. Suppose that circles B and C overlap completely. Is it possible to estimate the covariance between A and B explanatory of C? Of course not. There is no variability left over in B once we consider its association with C. We will discover in chapter 9 that even a higher degree of collinearity, or overlap, among explanatory variables can cause problems in multiple linear regression models.
6. *Linearity*: The mean value of *y* for each specific combination of the *x*s is a linear function of the *x*s. For example, the regression surface is flat in three dimensions. Note that in a simple linear regression model, we had to assume only a straight-line relationship, but now that we have two or more explanatory variables, we must move to higher dimensions. One way to think about this is to imagine a 3-D space and then visualize the difference between a flat surface and a curved surface. We assume that the relationship between the *x*s and *y* is not curved for any *x*. In chapter 10, we will find out about some of the tools for analyzing relationships that are not linear.
7. *Homoscedasticity*: The variance of the error term is constant for all combinations of *x*s. The term *homoscedasticity* means "same scatter." Its antonym is *heteroscedasticity* ("different scatter"). We will learn much more about this issue in chapter 11.
8. *Autocorrelation*: There is not an autocorrelated pattern among the errors. This is a huge problem in times series and spatial regression models. These involve data collected over time, such as months or years, or across spatial units, such as counties, states, or nations. Another name for this problem in over-time data is serial correlation. The key issue that leads to autocorrelation is that errors, or residuals, that are closer together in time (year 1 and year 2) or space (Los Angeles County, California, and Orange County, California) share more unexplained variance than errors that are farther apart in time (year 1 and year 10) or space (Los Angeles County, California, and Fairfax County, Virginia). Chapter 11 includes a discussion of autocorrelation.

Although not normally listed as an assumption, we also assume that there are not any wildly discrepant values that affect the model. This is related to assumptions about measurement error, specification error, and linearity. Suppose, for instance, that you look at a variable such as annual income and find the following values: [$25,000; $35,000; $50,000; $75,000; $33,000,000]. As we will see in chapter 12, the last entry is labeled as either a leverage point (if income is an explanatory variable) or an outlier (if income is the outcome variable). In either case, these types of discrepant values—which are known generally as *influential observations*—can affect, sometimes in untoward ways, linear regression models. The question you should ask when confronted by such values is: Why have they occurred? Did someone record the wrong value, perhaps by placing a decimal place in the wrong spot? Or is this value accurate; does someone in your sample actually earn that much money per year? If the value is simply coding error, then fixing it is usually simple. If it is an actual value, then perhaps a linear model predicting income, or using it as a predictor, is not appropriate. In this case, we should search for other regression techniques to model income. Fortunately, multiple linear regression models are flexible enough to accommodate many such problems; you just need to know where to look for solutions.

There are two general issues to consider: First, how do we test these assumptions; and, second, what do we do if they are violated? There are tests—most of them indirect—for all of these assumptions. Furthermore, if they are violated—if our model does not meet one or more of these assumptions—there usually are solutions that still make linear regression models viable. Several of the subsequent chapters in this book discuss these tests and the solutions. The term *regression diagnostics* is used as an umbrella term for these tests.

Several of these assumptions are stringent, but, as one argument goes, our models are saved by the *central limit theorem* (CLT) even when one or more of these assumptions is violated. You might recall from introductory statistics that the CLT states that for relatively large samples, the sampling distribution of the mean of a variable is approximately normally distributed even if the distribution of the underlying variable is not normally distributed (for a review, see Weiss, 1999). A bit more formally, we state: For random variables with finite sample variance, the sampling distribution of the standardized sample means approaches the standard normal distribution as the sample size approaches infinity.

This concept concerns the sampling distribution of the mean, which, as we have seen before, assumes that we take many samples from the population. If these samples are drawn randomly, the distribution of means tends to approximate the normal distribution after only about 30 samples. However, if the underlying variable's distribution is highly skewed, it may take more samples to approximate the normal distribution. Because intercepts and slopes are similar to means, they also may be shown to follow particular normal-like distributions (such as the *t* distribution). Therefore, given a large enough sample, we can infer that even if the outcome or explanatory variables are not normally distributed, the results of the regression model will be accurate most of the time. How large a sample needs to be to be large enough is a difficult question to answer, though. Thus, it is important to learn techniques appropriate for when assumptions like linearity or normality are not met, and this discussion should also reinforce the idea that using random samples or having control over explanatory variables is important.

Let's assume, though, that the assumptions are met. Then, according to something known as the Gauss–Markov theorem, the OLS estimators for the slopes, intercepts, and standard errors offer the best linear unbiased estimators among the class of linear estimators. No other linear estimator has less bias, or is more accurate on average, than OLS estimators.

Some Important Aspects of Multiple Linear Regression Models

As mentioned earlier in the chapter, OLS is the main technique that we use to estimate a multiple linear regression model. As we've now learned, OLS has some nice aspects that make it especially useful in regression analysis. These aspects are important because the OLS regression equation ($\hat{y}_i = \hat{\alpha} + \hat{\beta}_1 x_{1i} + \hat{\beta}_2 x_{2i} + \cdots + \hat{\beta}_k x_{ki}$) results in what is known as a *linear combination* of the *x*s that has the largest possible correlation with the *y* variable. This is an ideal because we are attempting, with this model, to explain as much of the variability in the outcome variable as possible with the explanatory variables. We will learn in chapter 4 about some ways to estimate this correlation.

We also mentioned in the last chapter that statistical software uses matrix routines to come up with slopes, standard errors, and other particulars of linear regression models. For those of you familiar with vectors and matrices, think of the *y* values as a vector of observations and the *x* values as a matrix of observations:

$$\mathbf{Y} = \begin{bmatrix} y_1 \\ y_2 \\ \vdots \\ y_n \end{bmatrix} \quad \mathbf{X} = \begin{bmatrix} 1 & x_{11} & \cdots & x_{1p} \\ 1 & x_{21} & \cdots & x_{2p} \\ \vdots & \vdots & \cdots & \vdots \\ 1 & x_{n1} & \cdots & x_{np} \end{bmatrix}$$

Given this shorthand way of expressing the explanatory and outcome variables (translate the vector and matrix into spreadsheet format, such as the data view in SPSS or Stata, if it's not clear how this works, and it is easy to see the utility of representing data this way), we may express the multiple linear regression equation as $\hat{\mathbf{Y}} = \mathbf{X}\hat{\beta}$, where the $\hat{\beta}$ represents a vector of the intercept (denoted in the **X** matrix with ones) and slopes for each explanatory variable. The **X** is listed first in this equation because we say it is *postmultiplied* by the vector of slopes.

Given the **X** and **Y** matrices, the following matrix formula may be used to estimate the vector of slopes:

$$\hat{\beta} = (\mathbf{X}'\mathbf{X})^{-1}\mathbf{X}'\mathbf{Y}$$

The **X** with the accent next to it is known as the *transpose* of the matrix, and the superscripted −1 indicates that we should take the inverse of the product in parentheses. In matrix terminology, we say the vector of slopes is estimated by postmultiplying the inverse of **X** times its transpose by **X**-transpose times **Y**. Try it out with a very small data set (if you have sufficient time and patience to compute the inverse), and see whether your results match what Stata comes up with. Matrix algebra is also useful for estimating several other features of multiple linear regression models. For example, the standard errors of the coefficients may be estimated by taking the square roots of the diagonal elements of the following matrix:

$$\mathbf{V} = (\mathbf{X}'\mathbf{X})^{-1}\sigma^2 \quad \text{where,} \quad \sigma^2 = \frac{1}{n-k} \times \hat{\varepsilon}'\hat{\varepsilon}$$

The term n refers to the sample size, and the term k refers to the number of explanatory variables in the model. We will learn more about σ^2 in chapter 4 when we discuss goodness-of-fit statistics. For now, note that it measures the amount of variability of the residuals around the regression line.

An alternative way of computing the standard errors that does not use matrix algebra is:

$$SE(\hat{\beta}_i) = \sqrt{\frac{\sum (y_i - \hat{y}_i)^2}{\sum (x_i - \bar{x})^2 \left(1 - R_i^2\right)(n-k-1)}}$$

We will learn more about the R^2 value in chapter 4. For now, we will say only that it is the R^2 from a linear regression model that includes x_i as the outcome variable and all the other explanatory variables as predictors (for example, $x_{1i} = \hat{\alpha} + \hat{\beta}_2 x_{2i} + \cdots \hat{\beta}_k x_{ki}$). This model is known as an *auxiliary regression model*. The quantity in parentheses that involves the R^2 is called the *tolerance*. Moreover, notice that the standard error is affected by the sum of squares of x, with larger values tending to decrease the standard error, and the sum of squared errors, with larger values tending to increase the standard error. A larger sample size also tends to decrease the standard error.

Kleinbaum, Kupper, Azhar, and Rosenberg (1998) provided a relatively painless overview of matrix routines useful in linear regression analysis. A more painful review, but one that is worth the effort for understanding the role of matrix routines in statistics, is found in Schott (1997). Unlike SPSS, Stata has a number of matrix routines built into its standard programs. Type `help matrix` in the command line to see its many options.

Key Terms

The following is a list of key terms and concepts in chapter 3. You can find definitions to these words throughout the chapter.

- Autocorrelation
- Best linear unbiased estimate (BLUE)
- Beta weights
- Central limit theorem (CLT)
- Collinearity
- Confound
- Dependent variable
- Distribution of error term
- Homoscedasticity
- Independence
- Independent variable
- Linearity
- Measurement error
- Specification error
- Statistical control

Analysis Exercises

1. Using either SPSS or Stata and the GSS 2010 data, complete the following tasks: Run a bivariate regression with days of poor mental health (mntlhlth) as the dependent variable and number of children (childs) as the independent variable. What association, if any, do you find between children and mental health?
2. Now, run a multiple regression with mental health as the outcome, with number of children and respondent's age (age) as predictors. What happens to the coefficient for *childs*? What is the relationship between age and poor mental health?
3. Now, add to the prior regression another predictor, feelings toward family income (finrela). What happens to the coefficients for age and *childs*? What effect does attitudes toward income have?

Solutions

1. Childs has no statistically significant association with number of poor mental health days in the bivariate model.

2. When we add respondent's age to the model, we find that the effect of childs is now significant, suggesting the importance of age as a control variable. Each additional child is associated with a 0.309-day increase in the number of poor mental health days. The effect is significant at $p < .05$. Meanwhile, each additional year of age is associated with 0.066-day decrease in poor mental health days.
3. When we add perceptions of income to the model, we find that the effect of childs is no longer significant and that the magnitude and significance of age decreases. However, more positive attitudes about income are associated with decreased mental health days.

CHAPTER 4 THE ANOVA TABLE AND GOODNESS-OF-FIT

Chapter Objectives

In chapter 3, we discussed what regression is and how to interpret its results. This chapter covers how to know whether your regression model has good fit to the data.

In chapter 4, we will

- further discuss how regression models are fit
- walk through the ANOVA table in SPSS and Stata
- discuss what makes a well-fit regression model.

As we discussed earlier, there are some similarities between regression analysis and analysis of variance (ANOVA) models. Both are useful, for example, if we wish to estimate the means of specific groups represented by the independent variables. Both models provide significance tests to determine whether there is a statistically significant difference between the mean *y* values for two or more groups signified by the *x* values. However, reach into your memory to recall the fundamental purpose of ANOVA. This is to partition the variance of a variable into component parts, one due to its systematic association with another variable and the other due to random error.

You probably noticed when you asked Stata or SPSS (depending on which one you are using) to estimate the model predicting number of poor mental health days in chapter 3 that you got a table that reported the results found in Table 4.1. This table lists the results of what is called the ANOVA. Why? Let's look at this example and then we can answer this question.

Sum of Squares

As in a standard ANOVA, there is a set of numbers known as the sum of squares. Recall from chapter 1 that we discussed a measure of dispersion known as the sum of squares of X, abbreviated SS[X]. The formula for this measure is $SS[X] = \Sigma(x_i - \bar{x})^2$. However, we now have an outcome variable, labeled *y*, and, as seen in the last chapter, we

Table 4.1: Analysis of Variance Table for Regression Model

Name in Stata	Name in SPSS	Value	*df*	MS (mean square)
Model	Regression	1,321.41	2	660.7
Residual	Residual	60,159.75	1145	52.54
Total	Total	61,481.16	1147	53.61
R-squared	R Square	0.021		
Adj R-squared	Adjusted R Square	0.019		
Root MSE	Std. Error of the Estimate	7.25		

wish to assess something about the correlation, or the overlap, between the explanatory and outcome variables, so it makes sense to try to come up with a measure of this overlap. Now, if the sum of squares of the outcome variable measures its total area, we might also wish for a measure of the overlapping area. This is what the ANOVA table provides. Under the table cell, which SPSS calls *Total Sum of Squares* (Total SS or TSS), we have SS[y]. Note that, again, there are slight differences in estimation between Stata and SPSS, but substantively, they give very similar results. For our purposes, we will use the Stata results. If we were to take all 1,148 values of the mental health variable, subtract the mean from each, and add these numbers up, we would obtain an overall sum of squares of 61,481.16. This may be thought of as the total area of a circle representing the variation in days of poor mental health. What, then, are those other sums of squares? The first one, in the *Regression Sum of Squares* (RSS) cell, or what Stata calls the Model SS, computes the variation of the mean of y around the values of y predicted from the regression equation. Recall that the predicted value is designated as $\hat{y}$ and is calculated based on the linear regression equation. Each nonmissing observation in the data set has a predicted value. We may, therefore, modify the sum of squares formula to reflect this type of variation:

$$\text{RSS} = \sum(\hat{y}_i - \bar{y})^2$$

This represents the total area of overlap between the variability of the xs and the variability of y (because the values of y are predicted based on the xs). It may also be thought of as the improvement in prediction over the mean with information from the explanatory variables. Its value in the linear regression model is 1,321.41.

The other sum of squares, in the *Residual Sum of Squares* (Residual SS) cell (better known as the sum of squared errors, or SSE; see chapter 2), is the area of y that is left

Figure 4.1: Visualizing the Analysis of Variance of a Linear Regression Model

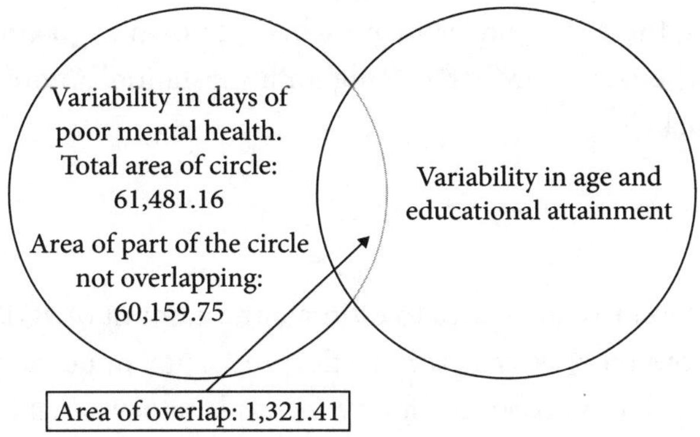

Figure 4.2: The Sum of Squares Equation

$$\sum(y_i - \bar{y})^2 = \sum(\hat{y}_i - \bar{y})^2 + \sum(y_i - \hat{y}_i)^2$$

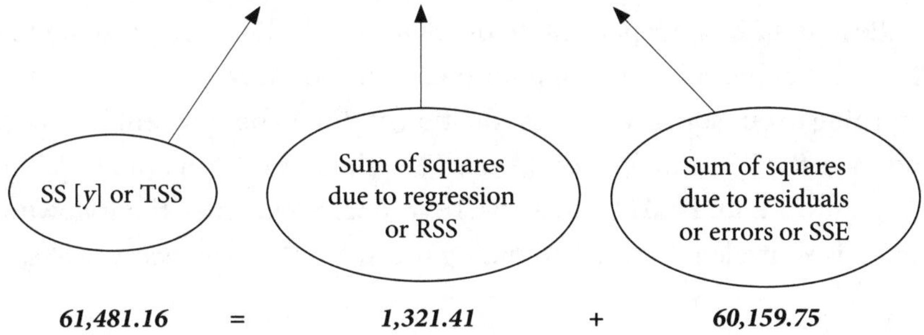

over after accounting for its overlap with the independent variables. It is computed using the following equation:

$$\text{SEE} = \sum(y_i - \hat{y}_i)^2$$

Its value in the Stata table is 60,159.75. Notice that the RSS and the SSE sum to equal the TSS: 1,321.41 + 60,159.75 = 61,481.16. This is because we assume we have accounted for all of the variation in *y* with either the variability in the *x*s or random variability (this does not, however, rule out that there might be other variables that account for portions of the variability of *y*). Figure 4.1 shows a picture of what we are examining with the ANOVA portion of our linear regression model of poor mental health days.

Another way of representing the equations inherent in the ANOVA table is to combine the sums of squares equations into one general equation. Such an equation should represent the partitioning of the area of the left-hand circle, or the TSS of violent crimes, into its two component parts. Figure 4.2 illustrates this concept.

The next issue to address is whether we can use this information on overlapping or joint variability to come up with a measure of the overall fit of the model. In regression analyses of all types, there is a concern with what is known as *goodness-of-fit* statistics. How do we know whether we are accounting for a substantial amount of the variability in the outcome variable?

R^2

One idea is to use the sums of squares to estimate the amount of variability. Because we have area measures of a type, we can estimate the proportion or percentage of variation in the outcome variable that is accounted for by the explanatory variables. There is a number labeled *R Square* next to the Stata table (or in SPSS in the model summary table); this is typically labeled simply R^2. Some people refer to it as the *coefficient of determination* or the *squared multiple correlation coefficient;* it indicates the proportion of variability in the outcome variable that is accounted for (some use the phrase "explained by") by the explanatory variables. It is simple to transform it into a percentage. In our example model, the R^2 is .022, or 2.2 percent. In other words, the linear regression model accounts for 2.2 percent of the variability in poor mental health days.

Returning to our sums of squares equation, it should be apparent how to compute this R^2 value: Take the RSS and divide it by the TSS. Therefore, from our model that predicted violent crimes, we find 1,321.14/61,481.16 = .022. Sometimes the following formula is used, but it is equivalent to simply dividing the RSS by the TSS because SSE + RSS = TSS:

$$R^2 = 1 - \frac{\text{SEE}}{\text{TSS}}$$

An important property of the R^2 measure is that it falls between zero and one, with zero indicating that the RSS is zero and a one indicating that the RSS = TSS, with the SSE equal to zero. More generally, this makes sense because we will explain somewhere between 0 percent and 100 percent of the variation in the outcome with our predictors.

An interesting phenomenon occurs when we add explanatory variables to the model: The R^2 increases whether or not the explanatory variables add anything of substance to the model's ability to predict *y*. Many researchers are understandably bothered by this problem as they do not wish to be misled into thinking that their model provides good predictive power when it does not. Thus, most use a measure known as the *adjusted* R^2. The term *adjusted* is used to indicate that the R^2 is adjusted for the number of explanatory variables in the model. If we simply add useless explanatory variables, the adjusted R^2 can actually decrease. The formula for this statistic is:

$$\bar{R}^2 = \left(R^2 - \frac{k}{n-1}\right)\left(\frac{n-1}{n-k-1}\right)$$

In this equation, as before, n indicates the sample size, and k indicates the number of explanatory variables in the regression model.

Another term next to the table is labeled Root MSE or *root mean square error*. It is not used very often in the social and behavioral sciences (although there are good arguments for using it), yet this standard error, which is sometimes referred to as the *standard error of the estimate* or the *standard error of the regression* (S_E or $S_{y|x}$), measures the average size of the residuals. It is also the square root of what is commonly referred to as the *mean square error* (MSE):

$$\text{MSE} = \frac{\text{SSE}}{n-k-1} \qquad \sqrt{\text{MSE}} = \text{S}_\text{E}$$

The MSE is found in the ANOVA table (in Stata it is called the Root MSE, whereas it is the Standard Error of the Estimate in SPSS). It may be used as a goodness-of-fit measure because smaller values indicate that there is less variation in the residuals of the model. It is also useful for estimating what are known as *prediction intervals*: bounds for the predictions we make with the model for particular observations within the data set. For instance, say that we wish to determine the number of poor mental health days for individuals with 12, 14, and 16 years of education. The following equation may be used:

$$\text{PI} = \bar{y} + \hat{\beta}_1(x_0 - \bar{x}) \pm (t_{n-2})(\text{S}_\text{E})\sqrt{1 + \frac{1}{n} + \frac{(x_0 - \bar{x})^2}{(n-1)\text{Var}(x)}}$$

A value such as 12 is substituted for x_0. The t value is the confidence level we wish to use with degrees of freedom equal to $n - 2$. For example, with a sample size of 50, we would use a t value of 2.011 for a confidence level of 95 percent (two-tailed). As with the confidence intervals for the slopes and the residuals from the model, both Stata and SPSS will compute prediction intervals (in Stata, use the predict command; in SPSS, see the Save window in the *Analyze-Regression-Linear* drop-down menu).

As with other statistics derived from a sample, the R^2 has a sampling distribution. If we were to take many samples from the population, then we could compute R^2s for each sample. We may, therefore, use significance tests and confidence intervals for this statistic. Before seeing such a test, though, we have to determine how R^2 values are distributed. We have, thus far, mentioned the normal distribution, the standard normal distribution (z distribution), and the t distribution. R^2 values follow an F distribution. Perhaps you remember the F distribution from the ANOVA model. If so, you will recall that it has two degrees of freedom, known as the numerator and denominator degrees of freedom. Notice that in the ANOVA table there are columns labeled *df* (degrees of freedom) and *MS* (mean square). The cells in the degrees of freedom column include the number of explanatory variables (k; in the Regression row), the sample size minus the number of explanatory variables minus one ($n - k - 1$; in the Residual row), and the sample size minus one ($n - 1$; in the Total row). The cells in the mean square column are the sums of squares in the rows divided by their accompanying degrees of freedom. Hence, based on the Stata output, the mean square due to regression is 1,321.41/2 = 660.70, and the mean square due to residuals is 60,159.75/1,145 = 52.54. This latter value is also the aforementioned MSE.

The number labeled F includes the *F* value, which is the mean square due to regression divided by the mean square due to residuals (MSR/MSE). Similar to the way we computed the *t* values for the slopes, this statistic may be thought of as a measure of how far our regression predictions are from zero, or how well our model is doing in predicting the outcome variable. The degrees of freedom for the *F* value are found in the df column: k and $(n - k - 1)$ or next to the *F* in parentheses. If we have a program that will compute it, such as Stata, we will rarely have to consult the *F* values in a table at the back of a statistics book or on the Internet. Rather, we find under the *F* value a number labeled Prob > *F* that contains the *p* value we need. This is the significance test for whether our model is predicting the outcome variable better with information from the explanatory variables than with simply the mean of the outcome variable. A better way, perhaps, of thinking about the *F* test (*F* value and its *p* value) is that it compares the null hypothesis that R^2 equals zero versus the alternative hypothesis that R^2 is greater than zero. If we find a significant *F* value, we gain confidence that we are predicting a statistically significant amount of the variability in the outcome variable. It should, by now, be rather unsurprising that we may also compute confidence intervals for the estimated R^2. However, this is done so rarely that we will not focus on it here.

Most researchers from the social and behavioral sciences use the R^2 and accompanying *F* test to determine something about model fit, or the importance of the model. You will often find statements such as "the model accounts for 42 percent of the variability in the tendency to twitch, and this proportion is significantly greater than zero." However, you should be cautious about such conclusions. Some statisticians argue that we should rely on either the substantive conclusions recommended by the partial regression slopes or the MSE (or S_E) to judge how well the set of explanatory variables predicts the outcome variable (see Graybill & Iyer, 1994, for a good discussion).

An Example of a Multiple Linear Regression Model

Before completing our discussion of the ANOVA table and goodness-of-fit statistics, let's see another example of a multiple linear regression model. We are going to use the same GSS data that we have used in previous chapters, this time with poor mental health days (mntlhlth) as the outcome and poor physical health days (physhlth), years at the same job (yearsjob), and income (income) as predictors. In Stata, we are going to

Table 4.2: Regression of Poor Mental Health Regressed on Independent Variables

Variable	*b*	*SE*	β
physhlth	0.300	0.032***	0.267
yearsjob	−0.030	0.018	−0.048
income	0.014	0.018	0.023
Constant	3.012	0.350***	
N	1,146		
R^2	.07		

***$p < .001$.

use the beta option, whereas in SPSS we will run the regression as normal. The results appear in Table 4.2.

First, let's look at the Model Summary information. The R^2 value is .07, the adjusted R^2 value is .067, and the Root MSE is 6.966 (Stata estimates). Hence, we may claim that, using the three explanatory variables, the model accounts for approximately 7 percent of the variability in poor mental health days. The output also indicates that the R^2 value is statistically distinct from zero, with an F value of 26.37 ($df = 3$, 1052) that is accompanied by a p value of less than .001.

The table of coefficients shows several results. It is best to ignore the intercept in this model because it is a rare person who has no income, has no days of poor physical health, and does not work. Another important issue here is that the three variables are measured on different metrics: physical health in days, years on the job in years, and income in dollars. Because of this, we cannot compare the coefficients they produce because they are not standardized. So how might we compare the coefficients to see which variable has the strongest effect on the outcome? In this case, there is only one statistically significant effect, but there is another statistical concept that can help us as well. The beta coefficients (beta weights), which measure the associations in standard deviation units, show that physical health has the strongest effect—a one standard deviation increase in poor physical health days is associated with a 0.25 standard deviation increase in poor mental health days. Here is one hint when reading beta coefficients: We care about the absolute value when comparing effects. For example, a beta coefficient of −0.45 is larger in magnitude than a coefficient of 0.15.

We must also consider the issue of sampling error: We do not know, based only on the point estimates, whether any two coefficients are significantly distinct. Perhaps the best method to do this is in Stata, by using the `test` postcommand. This allows a comparison of the coefficients to determine whether one is statistically distinct

Figure 4.3: Stata Output

```
.test physhlth = years job
 (1) physhlth – years job = 0
     F(1, 1052) = 58.80
         Prob > F = 0.0000
```

from another. For example, after estimating the regression model, we may type `test physhlth = yearsjob`. The result is shown in Figure 4.3.

This F test shows that there is a statistically significant difference between the two coefficients. Hence, we can conclude that physical health is a more substantial predictor of poor mental health than years on the job.

The coefficient for years on the job offers an interesting problem. The p value for this coefficient is .099, slightly above the threshold for making claims of statistical significance. Yet we could make a claim here that this result is statistically significant. It seems perfectly reasonable to create, a priori, a hypothesis that states that as years on the job increase, poor mental health decreases. If this is the case, then our hypothesis is directional and we may use a one-tailed significance test or confidence interval. As mentioned earlier, to switch from a two-tailed test (the default in both Stata and SPSS) to a one-tailed test, we divide the p value in half. Thus, the one-tailed p value is .099/2 = .0495. Voilà! We now have a significant result. This demonstrates the uncertainty involved in many statistical endeavors. Some might see this statistical sleight of hand as entirely justifiable, whereas others may argue that we are being disingenuous. There is no simple answer, though, so we should be both *careful and transparent* when setting up our hypotheses and describing the types of significance tests we are using when estimating a regression model.

Key Terms

The following is a list of key terms and concepts in chapter 4. You can find definitions to these words throughout the chapter.

- Adjusted R^2
- Goodness-of-fit
- Mean square error
- Regression sum of squares (RSS)
- Residual sum of squares (SSE)
- R^2 (coefficient of determination or square multiple correlation coefficient)
- Standard error of the estimate or standard error of the regression

Analysis Exercises

For this set of exercises, we are going to return to the regression run in the chapter 3 exercises.

1. Using the bivariate regression between mntlhlth and childs, what are the RSS and SSE of the regression? Now focus on the R^2 of the regression. How would you interpret the R^2 measure?
2. Run the multiple regression with mental health as the dependent variable and both number of children and age as predictors. What happens to the RSS, SSE, and R^2 compared with the bivariate model run in question 1.
3. Now add the family income measure (finrela) to the model. Compare R^2 and adjusted R^2. What differences are there between the two? What explains the size of the difference between the two?

Solutions

1. The RSS for the model is 12.62, while the SSE is 53.55. The R^2 is .0002, meaning that number of children accounts for only 0.02 percent of the variation in poor mental health days.
2. The RSS has increased to 401.56, while the SSE has decreased to 52.92 with the addition of age to the model. Our R^2 has also increased to .013, meaning that 1.3 percent of the variation in poor mental health days is accounted for by the independent variables.
3. The R^2 is .032, while the adjusted R^2 is .029. This is not a large difference, probably due to the fact that we have a relatively large sample size (n = 1,140) and added only one predictor to the model from that run in question 2.

CHAPTER 5 COMPARING LINEAR REGRESSION MODELS

Chapter Objectives

This chapter covers comparing two linear regression models to one another. Researchers often do this to see whether the addition of a variable or variables improves model fit.

In chapter 5, we will

- discuss why the addition of variables can improve model fit
- describe the use of the *F* test for comparing two models
- briefly go over confounding variables and their influence on regression results.

Now that we know something about ANOVA tables and goodness-of-fit statistics, we might be tempted to claim that it is a simple matter to use this information to compare different linear regression models. For example, suppose we take the model in the last chapter that sought to predict poor mental health days and add another variable to the model. How do we determine whether the new model, to use a common phrase, "fits the data better" than the initial model? Perhaps your reply is "that's easy; we simply look to see whether the additional variable is a significant predictor of the outcome variable," or "I'll look to see whether the R^2 has increased from one model to the next." Although this approach might work, it often will not. If we decide that all three variables in the first model are significant predictors, then simply adding another variable and checking whether its coefficient is significant could be problematic.

Regression models can be funny things, and you may find that various differences arise when you begin to add and take away variables from a model. We will discuss this issue in greater detail in chapter 7. For now, we will discuss model fitting procedures, mainly by comparing models with different variables. First, we need to introduce a couple of terms. A *nested model* is one that has the same subsets of variables but fewer overall explanatory variables than another model. For example, suppose we have two regression models designed to predict serum obesity levels among a sample of adult males. The first model includes the following explanatory variables:

{education, exercise frequency, age}.

The second model includes the following explanatory variables:

{education, exercise frequency, age, race/ethnicity, smoking status}.

We say that the first model is *nested* within the second model because it includes a subset of the variables. Sometimes researchers will call a nested model *hierarchical.* However, this can be slightly confusing, because a commonly used method in the social sciences is hierarchical (or multilevel) linear modeling, which has nothing to do with a nested model. Often, the model with the additional explanatory variables is called the *full* model; the nested model is sometimes referred to as the *restricted* model.

Nonnested models are two (or more) regression models that do not include the same subset of variables. For example, if the first model included education, exercise frequency, and age as explanatory variables, and the second included race/ethnicity, smoking status, and time spent each day watching television, we cannot say that one model is nested within another. Because comparing models in the nonnested situation is difficult, we will not pursue this issue here.

The Partial *F* Test and Multiple Partial *F* Test

Nested models are relatively easy to deal with. Let's begin with the simplest scenario: We add one more explanatory variable to the model and see whether it does a better job predicting the outcome variable. The two general tests that we may use are a t test and a type of F test known as a *partial* F *test*. The partial F test relates the extra sum of squares attributable to adding one explanatory variable to the model to the mean square error (MSE) from the full model:

$$F = \frac{\text{RSS(full)} - \text{RSS(restricted)}}{\text{MSE(full)}} \qquad df = 1,\ n-k-1\,\text{(full model)}.$$

One of the advantages of this test is that, as we shall see, it generalizes well when we wish to add more than one explanatory variable to the model. However, it may be inefficient in the current situation because we need to check a table of F values to determine the p value. As an alternative, most researchers use the t test. The t value from a two-tailed t test is the square of the F value from the partial F test. Moreover, the t test is found in the Stata printout: It is the t value and accompanying p value for the extra explanatory variable that has been added to the model.

Table 5.1: Ordinary Least Squares Regression for Religious Attendance

Variable	*b*	*SE*
educ	−0.009	0.013
inc10k	−0.026	0.022
age	0.022	0.003***
Intercept	2.619	0.251
N	2,033	
R^2	.021	

***$p < .001$.

However, suppose we wish to compare two models, one of which is nested in the other, but the full model contains two or more extra explanatory variables. In this situation, we extend the partial *F* test by using the *multiple partial* F *test.* As the formula for this test shows, it is a logical extension of the partial *F* test:

$$F = \frac{[\text{RSS(full)} - \text{RSS(restricted)}]/q}{\text{MSE(full)}} \qquad df = q,\ n - k - 1\,(\text{full model})$$

In this equation, we compute an *F* value based on the difference between the regression sums of squares RSSs from the two models. However, note that this difference is divided by *q*—the difference in the number of explanatory variables between the two models [(*k*(full) – *k*(restricted)]—before dividing by the mean square error of the full model. The degrees of freedom for the *F* value are *q* and the degrees of freedom associated with the error sum of squares for the full model.

As an example, let's return to the GSS 2010 data set. In previous chapters, we focused on days of poor mental and/or physical health, but we are going to change our example in this chapter. We are going to predict religious attendance (*attend*) in our model, with the predictors of education (*educ*), income in $10,000 increments (*inc 10k*) and age. As you have likely become familiar with Stata and SPSS output at this juncture, we provide the result of the restricted (or nested) multiple linear regression model in Table 5.1. Note, however, that SPSS and Stata can sometimes come up with slightly different results because of small differences in algorithms and the way that these programs make their calculations. These results indicate that these explanatory variables account for about 2 percent of the variability in religious attendance. Moreover, there is a positive association between age and religious attendance: Older people tend to attend church more frequently.

Now, let's see what happens when we rerun the model but add number of children to the model. Table 5.2 contains the results of the full model. It appears we have

Table 5.2: Ordinary Least Squares Regression for Religious Attendance with Childs Variable

Variable	*b*	*SE*
educ	0.007	0.013
inc10k	−0.023	0.021
age	0.012	0.004***
childs	0.264	0.039***
Intercept	2.619	0.25
N	2,033	
R^2	.043	

***$p < .001$.

added some important information to the model. Number of children, in particular, has a highly significant association with religious attendance after controlling for the effects of the other variables in the model. In fact, if the beta weights can be trusted (and we cannot know without checking the distributions of all the explanatory variables whether they are suitable for comparing the associations), number of children has the strongest association with frequency of attending religious services among the four variables.

Can we also determine whether the addition of the two explanatory variables significantly improves the predictive power of the model? Perhaps it would be useful to compare the R^2 values. After all, they measure the proportion of variability in y that is explained by the model. The R^2s are .021 for the restricted model and .043 for the full model. Unfortunately, it is not a good idea to compare these statistics. Recall that, in chapter 4, we learned that the R^2 goes up simply by adding explanatory variables, whether they are good predictors or not. An alternative is to compare the adjusted R^2 values as these are not as affected by the addition of explanatory variables. The values are .020 for the restricted model and .041 for the full model. It appears that this is a substantial increase: The adjusted R^2 more than doubled, but is the increase in explained variance statistically significant? We may use the multiple partial F test to answer this question:

$$F = \frac{676.16 - 338.01}{7.44} = 44.97\,(df = 1,2027)$$

An F value of 22.73 ($df = 1$, 2027) has a p value of approximately .0001, which is much lower than the threshold recommended by most rules of thumb. In this situation, it is safe to conclude that the addition of the gross state product and migrations per 100,000 rate increases the explained variance in the model.

Figure 5.1: Stata Output

```
test childs
 (1) childs = 0
     F(1, 2027) = 44.97
         Prob > F = 0.0000
```

Stata provides a simple way to compute a multiple partial F test with its test post-command. After estimating the full model, type `test childs`. The Stata output is shown in Figure 5.1. This is simply the same approach that we calculated using the sums of squares and the MSE. Note what we are doing: testing whether the two coefficients are equal to zero. This is fundamentally what a nested F test examines.

In SPSS, nested models can be run and compared by using the drop-down menus *Analyze-Regression-Linear*. In the first model, place the dependent and independent variables from the full model in the appropriate boxes, and click Next above the independent variables. A new window will then appear, which is labeled Block 2 of 2. Here, you can place the variable or variables you want to remove from the full model and change the Method to Remove. Next, click Statistics and ask for the R^2 change, which will run the F test for the nested model.

However, is this all we would wish to do with this model? We added one variable that appears to be strongly associated with religious attendance. Of course, we may want to add other variables to the model. In these cases, we could run partial F tests. However, at some point—some argue it should be very early in the analysis process—you as the analyst must decide which variables are important and which variables are not. This should be guided, preferably, by your theory or conceptual model, rather than by rote estimation of regression models or correlation matrices.

Confounding Variables

There is one other issue to discuss about our multiple linear regression models. This returns us to a point made in chapter 3 about the need for explanatory variables to check for confounding. As mentioned, a confounding variable accounts for the association between an explanatory and an outcome variable. We will discuss confounding in detail in chapter 7, but for now, it is useful to consider the two regression models designed to predict religious attendance. Looking back over the coefficients from the two models, notice that in the restricted model, age has a partial slope of 0.023, whereas in the full model, it has a partial slope of 0.011. In other words, the partial slope decreased by about 50 percent when we included number of children in the model. We cannot tell at this point whether this decrease is statistically significant (although there are tools for this), but let's assume it is. We may then claim that the new explanatory variable included in the model confounded the association between age and religious attendance. It did not confound it completely—the slope for age would be statistically indistinct from zero if this was the case—but it did change it enough to draw our attention. We usually look for variables that completely account for the association between two variables, but even those that only partially account for it can be interesting. The key question you should always ask yourself, though, is why. Why does the association change when we add a new variable? It could be a random fluctuation in the data, but it might be something important and worthy of further exploration. In this example, perhaps age and number of children are associated.

Key Terms

The following is a list of key terms and concepts in chapter 5. You can find definitions to these words throughout the chapter.

- Confounding variable
- *F* test
- Full model
- Nested modeling
- Restricted model

Analysis Exercises

1. Run a multiple linear regression model with days of poor mental health (mntlhlth) as the outcome and number of children (childs) and family income (finrela).
2. Now add the variable for poor physical health days (physhlth) to the model from #1.
3. Run an *F* test to see whether the addition of physical health improved model fit.

Solutions

1. The results for the model are in Figure 5.2.
2. The results for the model are in Figure 5.3.

The value of the F test is 77.94, with $p < .001$, indicating that the addition of physical health into the model improves fit.

Figure 5.2: Stata Linear Regression Output

. reg mntlhlth childs finrela

Source	SS	df	MS
Model	1403.25651	2	701.628256
Residual	60020.7119	1138	52.7422776
Total	61423.9684	1140	53.8806741

Number of obs	=	1141
F(2, 1138)	=	13.30
Prob > F	=	0.0000
R-squared	=	0.0228
Adj R-squared	=	0.0211
Root MSE	=	7.2624

mntlhlth	Coef.	Std. Err.	t	P > |t|	[95% Conf.	Interval]
childs	.0148859	.1409806	0.11	0.916	−.2617252	.2914971
finrela	−1.292221	.251699	−5.13	0.000	−1.786067	−.7983743
_cons	7.567815	.811482	9.33	0.000	5.975646	9.159983

Figure 5.3: Stata Linear Regression Output with physhlth

. reg mntlhlth childs finrela physhlth

Source	SS	df	MS
Model	5271.79634	3	1757.26545
Residual	56091.1324	1132	49.5504703
Total	61362.9287	1135	54.0642544

Number of obs	=	1136
F(3, 1132)	=	35.46
Prob > F	=	0.0000
R-squared	=	0.0859
Adj R-squared	=	0.0835
Root MSE	=	7.0392

mntlhlth	Coef.	Std. Err.	t	P > |t|	[95% Conf.	Interval]
childs	−.0188384	.1369082	−0.14	0.891	−.2874608	.2497839
finrela	−1.111411	.2449811	−4.54	0.000	−1.592079	−.6307424
physhlth	.2830475	.0320601	8.83	0.000	.2201436	.3459515
_cons	6.260025	.8019932	7.81	0.000	4.686464	7.833585

CHAPTER 6 DUMMY VARIABLES IN LINEAR REGRESSION MODELS

Chapter Objectives

This chapter covers the use of dummy variables, sometimes called dichotomous variables, to explain how qualitative characteristics, such as gender or religious affiliation, affect an outcome.

In chapter 6, we will

- discuss what a dummy variable is
- explain how to include it in a regression analysis
- explain how to interpret dummy variables and multinomial dummy variables
- discuss predicted values with dummy variables
- interpret dummy variables in the presence of continuous independent variables.

The term *dummy variable* is not meant as an insult. We are not claiming that one of our variables is dimmer or duller or has a lower IQ than another. In fact, the term's origins come from mannequins or ventriloquists' puppets—often known as *dummies*—which represent or take the place of real people. Dummy variables are used to represent something of particular interest in the regression model, such as a treatment effect (for example, going through a program or not, receiving treatment) or a specific group in the sample (for example, high school graduates, married people). A dummy variable is defined as a type of variable that takes on two values so that different categories of a categorical variable may be represented in the regression model. If you have ever conducted an ANOVA to determine the mean differences in some outcome for two or more groups, you have probably seen dummy variables. Often, these are called *indicator variables* to designate that they indicate a group represented in the sample. It is important to remember that the groups represented by the dummy variables must be mutually exclusive.

Research in the social, behavioral, and health sciences is full of dummy variables. For instance, suppose we wish to determine whether gender or ethnicity is associated

Table 6.1: Example Data

Observation	Ethnicity	x_1	x_2	x_3
1	Caucasian	1	0	0
2	Caucasian	1	0	0
3	African American	0	1	0
4	African American	0	1	0
5	Hispanic	0	0	1
6	Hispanic	0	0	1

with some continuous outcome variable. Clearly, gender, ethnicity, religious group affiliation, marital status, family structure, and many other variables cannot be represented by continuous, let alone normally distributed, variables. Yet we often wish to include these types of items in regression models. Dummy variables provide an efficient way of including categorical or binary (two-category) variables. Here are some examples of the coding associated with dummy variables:

Gender: 0 = male; 1 = female

Ethnicity (Caucasian, African American, Hispanic): x_1: Caucasian = 1; other = 0; x_2: African American = 1; other = 0; x_3: Hispanic = 1; other = 0

The first variable, *gender*, is straightforward as we place people into one of only two groups. The second set of variables requires an explanation, though. Assume that we have a variable in our data set that is designed to measure ethnicity (we will limit this example to three ethnic groups for the sake of simplicity). It has three outcomes—Caucasian, African American, and Hispanic—that are coded 1, 2, and 3. Because this variable is not continuous, nor can it even be ordered in a reasonable way, we could not include it as is in a linear regression model (what would a one-unit increase or decrease mean?). The solution is to create three dummy variables that represent the three groups. The main group represented by each is coded as 1, with any sample member not in this group coded as 0. We can extend this form of dummy variable coding to a variable with any number of unique categories. Table 6.1 shows example data. The variables x_1, x_2, and x_3 are dummy variables that represent the three ethnic groups in the data set. There are several other types of coding strategies possible, such as effects coding and contrast coding, but the type shown in the table is flexible enough to accommodate many modeling needs. We could also add additional groups (for example, East Asian, Native American, Aleut) if our original variable included codes for them.

There is a rule you should always remember when using dummy variables to represent mutually exclusive groups (for example, married people, never-married people, and widowed people) in a regression model: When setting up the model, include all the dummy variables in the model except one. The one dummy variable that is omitted from the model is known as the *reference category* or *comparison group*. Suppose our variable has five categories (represented by k) and we create five dummy variables to represent these categories. We then use four (or $k - 1$) dummy variables in the regression model. If you try to use all five groups, the model will not run correctly because there is perfect collinearity among the dummy variables (we discuss collinearity in greater detail in chapter 9). How should you choose the reference category? This is up to you, but there are some guiding principles that may be helpful. First, many researchers use the most frequent category as the reference category; although, it is usually preferable to let your hypotheses guide the selection. Second, it is not a good idea to use a relatively small group as the reference category. Third, many statistical software programs have regression routines that automate the creation of dummy variables. When these are used, you should check the software documentation to see which dummy variable is excluded from the model. Some programs exclude the most frequent category; others exclude the highest or lowest numbered category.

How are dummy variables used in a regression model? Just like any other explanatory variable. It is often valuable to give the variables names that are easy to recognize. For example, if we were to create a set of dummy variables representing marital status, we might wish to name them *married*, *divorced*, *cohabiting*, *single*, and *widowed*.

Let's first consider a regression model with only one dummy variable. This dummy variable is labeled *sex*. The outcome variable is annual personal income in ten thousands of dollars (*income06*). If only gender is in the model, it is identical to an ANOVA model and, as we shall see, may be used to determine the mean income levels for males and females. Using the GSS 2010 data set, we are going to focus on the variable *sex*, which is coded as 1 = male and 2 = female. This illustrates a frequent coding situation: There are many data sets that code dummy variables as {1, 2}. Although we may still use these variables in a regression model, the {0, 1} coding scheme is much more preferable for reasons that will become clear later on. To reflect this more desirable coding scheme, we have created a variable in the data set called *gender,* where 0 = male and 1 = female. In addition, the data set includes a marital status variable that is nominal (*marital*) that could be made into dummy variables representing specific marital statuses; and a race/ethnicity variable (*hhrace*) along with a set of dummy variables (*Caucasian*, *AfricanAm*, *othrace*) that represent its categories. There are also several other variables in this data set that measure family income, religion, volunteer activities, and various demographic characteristics. (Note: The income variable we are using for this example is not truly continuous, nor are the variables coded correctly. Rather, they include categories of

Table 6.2: Simple Ordinary Least Squares Regression for Income

Variable	b	SE
Gender	−1.349	0.282***
Intercept	17.111	0.209
N	1,805	
R^2	.013	

***$p < .001$.

income that are not in actual $10,000 units, even though they are labeled as such. Nonetheless, in the interests of unfettered learning, we will treat them as continuous and pretend that the units are accurate.) Setting up a linear regression model with dummy variables in Stata or SPSS requires no special tools or unique approach. The setup is the same as with continuous explanatory variables: Simply place the dummy variable(s) in the regression command following the outcome variable. Let's run the model specified in the last paragraph and see what Stata and SPSS provide, as shown in Table 6.2.

The output looks the same as before, with coefficients, standard errors, t values, and p values. In fact, most of the interpretations are similar, except the most important: the coefficients. Recall that we have learned that the intercept is sometimes not very useful in linear regression models because explanatory variables are frequently coded so as not to have an interpretable zero value. Think about how a dummy variable is different: It clearly has an interpretable zero value since it is coded as $\{0, 1\}$. However, because it no longer makes much sense to refer to a one-unit increase in the explanatory variable—at least not in the same way as with continuous explanatory variables—we need to modify our thinking about the slope. Perhaps writing out the linear regression model will help us move toward a clearer interpretation of the results:

$$\text{Personal income, predicted} = 17.111 - 1.349\ (\text{gender})$$

Suppose we wish to compute predicted values from this equation. This should be pretty easy by now:

Predicted income for males:	$17.111 - 1.349(0) = 17.111$
Predicted income for females:	$17.111 - 1.349(1) = 15.762$

Consider that the mean of personal income is 16.367, which looks remarkably close to the average of these two predicted values. It should be for the simple reason that, as foreshadowed earlier, these two predicted values are the mean values for personal income among males and females, while the overall mean for the outcome variable is the weighted average of the two separate means by gender.

Given that we may predict these two means with the linear regression model, what do the intercept and slope represent? It should be apparent that the intercept, when there is only one {0, 1} coded dummy variable, is the mean of the outcome variable for the group represented by the zero category. The slope represents the average difference in the group means of the outcome variable. Hence, we may say that the expected difference in personal income between males and females is 1.349. Without even computing predicted values, we immediately see that females, on average, report less personal income than males.

An individual experienced with various types of statistical analysis might ask: Isn't this the type of issue that *t* tests are designed for? Of course it is. In fact, running a *t* test in either Stata or SPSS reveals the same results found in the linear regression model: an average difference between males and females of 1.349, a *t* value of 4.781, and a *p* value of less than .001.

Dummy Variables in Multiple Linear Regression Models

Considering that the two-sample *t* test provides the same information as the dummy variable approach in a simple linear regression model, are there any advantages to using dummy variables? The main advantage is that we may include additional explanatory variables (either continuous or dichotomous) in the regression model to come up with predicted means that are adjusted for continuous or other dummy variables. A linear regression model offers a straightforward way to estimate this type of model, but first let's see another nice property of the linear regression approach by adding some more dummy variables, except that this time we will add dummy variables to explore differences among several groups.

Let's return to the personal income model and add the race/ethnicity dummy variables. First, we need to decide which of this set to exclude as the reference category. The frequencies for the three groups are 1,549 (Caucasian), 290 (African American), and 198 (other race/ethnicity). We have not set up a hypothesis that guides the selection of the reference category, so it will be an arbitrary selection in our example. Let's choose the most frequent category—Caucasian—as our comparison group. We will then estimate a linear regression model (shown in Table 6.3) that includes *gender* but adds the two dummy variables that identify African Americans and members of other racial/ethnic groups to predict personal income.

Considering the way we have treated dummy variables so far, what do you think these slopes and the intercept measure? As before, let's write out the regression model:

$$\text{Predicted income} = 17.692 - 1.193\ (\text{gender}) - 4.012\ (\text{African American}) \\ - 0.902\ (\text{other race/ethnicity})$$

Table 6.3: Ordinary Least Squares Regression for Income with Gender and Race

Variable	*b*	*SE*
Gender	−1.193	0.276***
African American	−4.012	0.396***
Other race	−0.903	0.462
Intercept	17.692	
N	1,799	
R^2	.066	

***$p < .001$.

Table 6.4: Expected Means by Gender and Race/Ethnicity

Group	Computation	Expected Mean
Caucasian		
Males	17.692	17.692
Females	17.692 – 1.193	16.499
African American		
Males	17.692 – 4.012	13.680
Females	17.692 – 4.012 – 1.193	12.487
Other race/ethnicity		
Males	17.692 – 0.903	16.789
Females	17.692 – 0.903 – 1.193	15.596

Because the intercept in a multiple linear regression model is the expected value of the outcome variable when all the explanatory variables are zero, 17.692 represents the expected mean for Caucasian males. In other words, the intercept represents the mean for the omitted category from both sets of dummy variables: gender and race/ethnicity. As we found earlier, the slopes represent average differences among the groups. These are also known as *deviations from the mean.* Here, we are interested in computing deviations from the mean of the reference category (Caucasian males), but we must now combine coefficients to estimate the means for these different groups, as shown in Table 6.4.

We have estimated means for each of the six mutually exclusive groups defined by the dummy explanatory variables. However, we also need to consider the *t* value and *p* values in the regression output. As in the earlier model, the gender coefficient is statistically significant. Another way of thinking about this is that the gender *difference* is statistically significant. However, the coefficient for African Americans is significant, whereas the coefficient for other race is not. This means that we cannot say that the income differences among Caucasians and members of other racial/ethnic groups are substantial enough that we could conclude that they differ in the population of adults in the United States. However, we can say that such a difference exists between Caucasians and African Americans. Also, notice that the income difference between males and females in each racial/ethnic group is identical (1.193). Whereas this difference is statistically significant, the income difference between males across racial/ethnic groups and females across racial/ethnic groups is not statistically significant (for example, the difference between Caucasian females [16.499] and females of other races/ethnicities [15.596] is not statistically significant).

Linear Regression Models with Dummy and Continuous Variables

The last situation we will address in this chapter involves linear regression models that include both dummy and continuous explanatory variables. The well-known analysis of covariance (ANCOVA) model is the analogue to this type of regression model. ANCOVA models serve several useful purposes. The main advantage is that we may compare groups regarding some variable after adjusting for (that is, controlling for) their associations with some set of continuous variables. Recall in the last chapter that we discussed briefly the issue of confounding variables. Here is another example of this phenomenon. Suppose we are studying rates of colon cancer among cities in the United States. We draw a sample of cities and find that those in Florida, Arizona, and Texas have higher rates of colon cancer than cities in the Northeast, Midwest, and Northwest areas of the country (perhaps we have set up a series of dummy variables that indicate region of the country). Are there unique environmental hazards in warm-weather cities that affect the risk of colon cancer? We cannot tell without much more information about the environmental conditions in these cities. Nevertheless, we have not yet considered a key confounding variable that affects analyses of most rates of disease: age. It is likely that many warm-weather cities—especially in states like Florida and Arizona—have, on average, an older population than cities in colder climates. Age is thus associated with both colon cancer rates and region of the country. ANCOVA models are designed to correct or adjust for continuous variables, such as age, that act as confounders.

Multiple linear regression models may be used in a similar manner and are often much easier to interpret. If we are interested in analyzing differences in some outcomes

Table 6.5: Ordinary Least Squares Regression for Income on Gender and Religion

Variable	*b*	*SE*
Gender	−1.322	0.282***
Fundamental	−1.458	0.322***
Intercept	17.519	0.209
N	1,718	
R^2	.025	

***$p < .001$.

among groups defined by dummy variables, we should consider adjusting for the possible association with conceptually relevant continuous variables. Here is an example that is similar to our earlier dummy variable example but changes the outcome variable. There is a perception in some circles that fundamentalist or conservative Christians tend to come from poorer families than do other people. Hence, we may surmise—and this is admittedly a crude hypothesis—that family income among fundamentalist Christians is lower on average than in other families in the United States. Let's test this rough hypothesis using the GSS 2010 data. Table 6.5 gives the results of a linear regression model with personal income (*income06*) as the outcome variable and a dummy variable (*fundamental*, coded 0 = nonfundamentalist, 1 = fundamentalist Christian). The model also includes gender as an explanatory variable.

The model provides support for the hypothesis: On average, adults in the fundamentalist Christian category have 1.458 units lower income than nonfundamentalist adults, with 0.672. This coefficient is statistically significant at the $p < .001$ level, with 95 percent confidence intervals of [−2.090, −0.827]. Let's think about this association a bit more. Are there other variables that might account for this negative association? Even if you are not an expert on fundamentalist Christianity in the United States, it is still possible to come up with some informed ideas. Let's consider income. What accounts for differences in income in the United States? A prime candidate is education. If we were to read a bit about religion in the United States, we might come across literature that suggests that conservative or fundamentalist Christians tend to have less formal education than others (although this has changed in recent decades).

A reasonable step is to include a variable that assesses formal education to determine whether it affects the results found in the previous model. The GSS 2010 data set includes such a variable, *educ*, which measures the highest year of formal education completed by sample members. What happens when we place it in the model? Rerun the regression model in Stata or SPSS (we do it in Stata here) but include the education variable. Results are represented in Figure 6.1.

Figure 6.1: Stata Linear Regression Output

. reg icome06 gender fundamental educ

Source	SS	df	MS
Model	11169.6868	3	3723.22894
Residual	50159.7974	1712	29.2989471
Total	61329.4843	1715	35.7606322

Number of obs = 1716
F(3, 1712) = 127.08
Prob > F = 0.0000
R-squared = 0.1812
Adj R-squared = 0.1807
Root MSE = 5.4129

income06	Coef.	Std. Err.	t	P > |t|	[95% Conf.	Interval]
gender	1.504716	.2632773	5.72	0.000	.9883373	2.021096
fundamental	−.6165531	.2988212	−2.06	0.039	−1.202646	−.03046
educ	.7745791	.0426734	18.15	0.000	.6908815	.8582767
_cons	5.385417	.6263472	8.60	0.000	4.15693	6.613903

Note that the coefficient associated with the fundamentalist Christian has reduced in magnitude (the coefficient is smaller) and in statistical significance level (now significant at $p < .05$). However, given the literature on formal education among fundamentalist Christians in the United States, we should strongly suspect that education partially accounts for the association between considering oneself a fundamentalist Christian and family income. If we included even more variables, the statistical significance of being a fundamentalist Christian may disappear altogether.

How can we estimate means for different groups with this regression model? By now, we hope it apparent that writing out the implied prediction equation is useful:

$$\text{Predicted income} = 6.89 - 0.617\ (\text{fundamental}) - 1.505\ (\text{gender}) + 0.775\ (\text{educate})$$

We have already seen what to do with the dummy variables: Simply plug in a zero or a one and compute the predicted values, but what should we do with education and its coefficient? It is obvious that we should do something similar with this variable as we did with the dummy variables: Plug in some value, compute the products, and add them up. Suppose we think that putting a zero or a one for education is the way to go. What would these numbers represent? Looking over the coding of *educate*, perhaps by asking Stata or SPSS to calculate summary statistics, we find that it has a minimum value of zero and a maximum value of 20. However, notice that there is only one sample member who reported one year of education; the next value in the sequence is five. Therefore, placing a zero or a one in the equation is unwise as these values are not represented well in the data set. A value of 20 might be reasonable, but for various reasons, it is better to use an "average" value of education in the computations (for example, mean, median, or mode) or some other sensible value. For example, in the United States, a well-recognized

Table 6.6: Adjusted Means for Regression in Figure 6.1

Group	Computation	Adjusted Mean
Fundamentalist males	$6.89 - 0.617 + (0.77 \times 12)$	15.569
Fundamentalist females	$6.89 - 0.617 - 1.505 + (0.77 \times 12)$	14.064
Nonfundamentalist males	$6.89 + (0.77 \times 12)$	16.185
Nonfundamentalist females	$6.89 - 1.505 + (0.77 \times 12)$	14.680

Table 6.7: Adjusted Means without Education for Regression in Figure 6.1

Group	Computation	Adjusted Mean
Fundamentalist males	$6.89 - 0.617$	6.273
Fundamentalist females	$6.89 - 0.617 - 1.505$	4.768
Nonfundamentalist males	6.89	6.89
Nonfundamentalist females	$6.89 - 1.505$	5.385

educational milestone is graduation from high school. This is normally denoted as 12 years of formal education. Twelve years is also the modal category in the distribution of *educate*. Hence, there is justification for using 12 in our computations if we wish to estimate means for the categories of *fundamental* and *gender* that are adjusted for years of education. The computations can be found in Table 6.6.

The average estimated difference in family income between fundamentalist Christians and other adults is 0.617, which is substantially smaller than the difference between males and females. Nonetheless, it is instructive to compare the means after adjustment for education with the unadjusted means. These latter values, based on the first family income regression model, are shown in Table 6.7.

To estimate the difference between adjusted and unadjusted values, it is helpful to calculate the percentage difference. Now that we have made all these computations by hand (and calculator), it is helpful to know about a Stata postcommand that may be used to simplify things quite a bit (unfortunately, a similar command is not present in SPSS). The `margins` command following the regression model computes predicted values from regression models. The key is to tell Stata the values of the explanatory variables. For example, after estimating the full model, the following command will provide the predicted value for nonfundamentalist males with 12 years of education (the results can be seen in Figure 6.2): `margins, at(fundamental=0 gender=0 educ=12)`.

Look over the documentation for Stata's margins command. It shows that there are several ways to represent predicted values. For example, if we wish to see the four

Figure 6.2: Prediction Using Stata's Margins Command

```
.margins, at (fundamental = 0 gender = 0 educ = 12)
Adjusted prediction                          Number of obs    =    1716
Model VCE  : OLS

Expression  : Liner prediction, predict ( )
at          : gender       =        0
              fundamental =        0
              educ         =       12
```

	Margin	Delta-method Std. Err.	t	P>|t|	[95% Conf. Interval]
_cons	14.68037	.2115956	69.38	0.000	14.26535 15.09538

Figure 6.3: Expected Values Using Stata's Margins Command

```
.margins, over (fundamental gender) at (educ = 12)
Predictive margins                           Number of obs    =    1716
Model VCE    :  OLS

Expression   :  Liner prediction, predict ( )
over         :  fundamental gender
at           :  0.fundamental#0.gender
                   educ           =     12
                0.fundamental#1.gender
                   educ           =     12
                1.fundamental#0.gender
                   educ           =     12
                1.fundamental#1.gender
                   educ           =     12
```

	Margin	Delta-method Std. Err.	t	P > |t|	[95% Conf. Interval]	
fundamental#gender						
0 0	14.68037	.2115956	69.38	0.000	14.26535	15.09538
0 1	16.18508	.2215189	73.06	0.000	15.75061	16.61956
1 0	14.06381	.2772692	50.72	0.000	13.51999	14.60763
1 1	15.56853	.2971237	52.40	0.000	14.98577	16.15129

groups at 12 years of education, we may use the margins command in the following way: `margins, over(fundamental gender) at(educ=12)`(results are shown in Figure 6.3).

Let's consider one more dummy variable regression example. This time we will consider the association between marital status and family income. It is reasonable to hypothesize that married people have higher incomes, on average, than single people,

Table 6.8: Ordinary Least Squares Regression for Income

Variable	b	SE
widow	−5.347	0.498***
divsep	−4.306	0.343***
nmar	−5.571	0.309***
Intercept	19.233	0.194
N	1,805	
R^2	.182	

***$p < .001$.

whether they are never married, divorced, or widowed. Families with a married couple have at least the potential to earn two incomes; in fact, more than half of married couple families in the United States have two wage earners. Therefore, let's run a regression model to assess the association between marital status and family income. Marital status is represented by a set of four dummy variables: *married, widow, divsep* (divorced or separated), and *nmar* (never married). We will use *married* as the reference category.

The resulting model has an R^2 of .182, an adjusted R^2 of .181, and an F value of 133.55 ($df = 3, 1801$; $p < .001$). The regression coefficients (as shown in Table 6.8) or slopes support the hypothesis that married people report more family income than others. The average differences are four to five units with, it seems, never married people reporting the least family income.

However, think some more about variables that are associated with marital status and income. We have seen, for instance, that education is strongly associated with income, but it also may be associated with marital status. Another variable mentioned earlier is age. Could age be related to income? It is possible, because middle-aged people earn more on average than younger people. What about age's association with marital status? It makes sense to argue that there should be an association, especially when we have a never-married category. Never-married people tend to be younger, on average, than married people. Given these important issues, it is prudent to add education and age to the model. The regression table provided by Stata is Table 6.9.

The new model, which we may call the *full model* to distinguish it from the earlier *restricted model*, has an R^2 of .322, an adjusted R^2 of .320, and an F value of 170.41 ($df = 5, 1795$; $p < .001$). Although we will skip the computations, a nested F test indicates that we have significantly improved the predictive power of the model by adding age and education (that is, the R^2 has increased by a statistically significant amount). You should

Table 6.9: Ordinary Least Squares Regression for Income

Variable	*b*	*SE*
widow	−4.764	0.494***
divsep	−3.986	0.314***
nmar	−5.131	0.312***
educ	0.729	0.038***
age	0.013	0.009
Intercept	8.46	0.693
N	1,801	
R^2	.322	

***$p < .001$.

note that a nested *F* test is not the same as subtracting the *F* values from the two regression models. This latter procedure is not appropriate for comparing regression models.

Although we may have suspected that the results for marital status change when adding age and education to the model, there is actually little shift in the size of the coefficients. For example, the never-married coefficient changes from −5.57 to −5.13, a difference of about only 4 percent. Hence, we may conclude that the differences in family income associated with marital status are not accounted for (nor are they confounded) by age or education.

There is one other issue that we should consider before closing this chapter. As mentioned in chapter 3, many researchers use standardized coefficients to compare the associations within a model. For instance, because education and age are measured in different units, trying to compare their association with an outcome variable such as family income in a multiple linear regression model is difficult. By using the standardized coefficients, so the argument goes, education and age may be compared in standard deviation units. Although there is some merit to this approach, it is not possible when using dummy variables. A one standard deviation unit shift in a dummy variable makes no sense (unless it equals exactly one, which does not happen with {0, 1} dummy variables). Dummy variables can shift only from zero to one or from one to zero, so interpreting the standardized coefficients associated with dummy variables is not appropriate. For example, in Figure 6.4, the variable *gender* has a mean of 0.564 and a standard deviation of 0.496. In the regression model shown earlier (and in the Stata output shown in Figure 6.4), the standardized coefficient associated with *gender* is 0.125. So the mindless interpretation of this coefficient is that "Controlling for the effects of the other variables in the model, a one standard deviation unit increase in

Figure 6.4: Ordinary Least Squares Regression Output in Stata with Standardized Regression Coefficients

```
. reg income06 gender fundamental educ, beta

      Source |       SS           df       MS      Number of obs   =      1716
-------------+----------------------------------   F(3, 1712)      =    127.08
       Model |  11169.6868         3  3723.22894   Prob > F        =    0.0000
    Residual |  50159.7974      1712  29.2989471   R-squared       =    0.1821
-------------+----------------------------------   Adj R-squared   =    0.1807
       Total |  61329.4843      1715  35.7606322   Root MSE        =    5.4129

------------------------------------------------------------------------------
    income06 |      Coef.   Std. Err.      t    P>|t|                     Beta
-------------+----------------------------------------------------------------
      gender |   1.504716   .2632773     5.72   0.000                 .1251072
 fundamental |  -.6165531   .2988212    -2.06   0.039                -.0456826
        educ |   .7745791   .0426734    18.15   0.000                 .4017544
       _cons |   5.385417   .6263472     8.60   0.000                        .
------------------------------------------------------------------------------
```

gender is associated with a 0.125 standard deviation increase in family income." A one standard deviation unit increase in *gender* shifts, for instance, from 0 to 0.496. This is a nonsensical value; there is no *gender* value of 0.496, nor can there be.

Key Terms

The following is a list of key terms and concepts in chapter 6. You can find definitions to these words throughout the chapter.

- Dummy variable
- Indicator variable
- Reference category

Analysis Exercises

1. We want to know whether females have more poor mental health days than men. Run a bivariate regression between mntlhlth and gender.
2. What happens to the relationship between mntlhlth and gender when income06, educ, and race (AfricanAm and othrace) are included in the model?
3. Interpret the effect of AfricanAm and othrace on mntlhlth, remembering that the reference category is Caucasian. Are there significant differences between African Americans and individuals who identify with other racial groups? (Hint: You will need to create a variable for Caucasian to test this.)

Solutions

1. The regression analysis shows that women, on average, have 0.982 fewer poor mental health days than men.
2. The size of the gender coefficient gets larger. Now, women report 1.053 fewer poor mental health days than men.
3. There are no race differences between Caucasians and African Americans or individuals of other racial/ethnic groups. There is a significant difference between African Americans and individuals of another racial/ethnic identity. African Americans report 1.9 fewer poor mental health days than other racial/ethnic groups.

CHAPTER 7 SPECIFICATION ERRORS IN LINEAR REGRESSION MODELS

Chapter Objectives

This chapter discusses some of the common errors in regression analyses and how to detect them, and it begins a discussion on how to fix them.

In chapter 7, we will discuss

- overfitting regression models
- underfitting regression models
- endogeneity
- the virtual certainty of specification errors
- ways to test for these errors.

We discussed briefly in the last two chapters an important issue that concerns multiple linear regression models: We should carefully consider the variables we place in our models. In chapter 5, we showed that there are statistical tools that can help us determine whether significantly more of the variance of an outcome is explained with the presence of an additional explanatory variable. We have also learned about the importance of confounding variables. If we fail to consider variables that account, either partially or fully, for the association between our key explanatory variables and the outcome variable, we may make incorrect conclusions about our data. Both of these issues relate to the topic of this chapter: *specification errors*. A specification error is a mistake when deciding the correct model for predicting an outcome. When establishing the term *correct model*, we should consider several issues. First, is a linear model appropriate? Suppose we have the association shown in Figure 7.1 between an explanatory and an outcome variable.

A linear regression model is going to attempt, using the least squares formulas, to fit a straight line to this set of data, but it is obvious that the association between x and y is not linear; rather, it is curved or *curvilinear*. Estimating this association with a linear model is one example of specification error. We will discuss curvilinear associations in much more detail in chapter 10.

Figure 7.1: Visualizing Specification Error

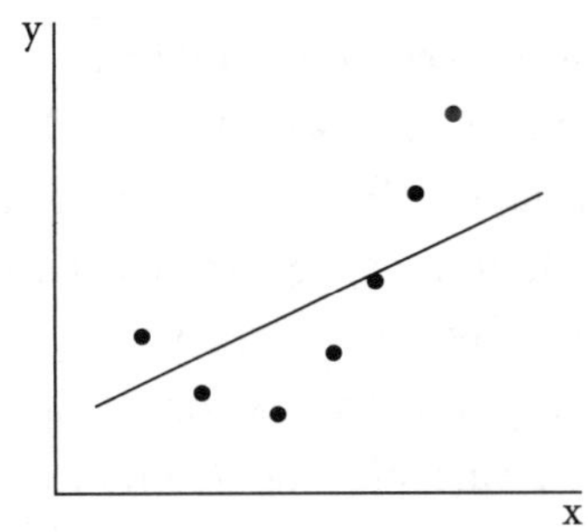

For now, we will learn about three common sources of specification error in multiple linear regression models: (1) including irrelevant variables; (2) leaving out important variables; and (3) misspecifying the "causal ordering" of variables within the model. The first problem is called *overfitting*; the second problem is called *underfitting*; and the third problem involves what is known as *endogeneity* or *simultaneous equations bias*. All three present problems for linear regression models, but to a different degree. Before learning about these problems, though, it is important to discuss for a moment variable selection.

How do we know which variables to put in a regression model? Variable selection is a frequently discussed topic in regression modeling. There are numerous variable selection procedures available to researchers. We will discuss some of these later in the chapter. At this point, it is important to realize that much of the work on variable selection by mathematical statisticians overemphasizes obtaining the best *prediction* equation at the expense of testing a reasonable *explanatory* model. Social scientists, including social workers, should be interested in producing a good explanatory model. In this way, we acknowledge that regression models are clearly useful as predictive tools. Most of us would prefer, for instance, that medical researchers use statistical tools to predict with a high degree of certainty that a drug will, at least, not injure us. Perhaps we do not care much about explaining the underlying biological mechanisms that link the drug with the cure. However, in the social and behavioral sciences, explanations are the driving force at the core of research. We wish to be able to explain why one variable is associated with another (for example, why are neighborhood unemployment rates associated with crime rates or poor health?) rather than to say only whether one predicts another. Therefore, when we select variables for a regression model, we should use our explanatory framework—our theory or conceptual model, with a healthy dose of researchers' intuition—to decide which variables to include in the regression equation. This presupposes that we have a solid understanding of the research literature, especially previous studies that have examined our outcome variable, and that we understand fully the variables in our data set, including how they are constructed, what they purport to measure, and so forth.

Overfitting—Or the Case of Irrelevant Variables

It is very easy to find examples of overfitted models in the literature. Browse through an issue of just about any journal in the social and behavioral sciences and you will see regression models with variables that add nothing to their ability to predict the outcome variable. Of course, there may be various reasons for including seemingly irrelevant variables in a model. Why would researchers include such variables in their models? Typically, they seek to guard against omitting potential confounding variables by including virtually any variable that seems to be important. This normally does not present a problem for linear regression models (except that it violates the sacrosanct principle of parsimony and can cause problems if we have a small sample size), but there are situations where we must be careful about putting too many variables in a model.

Overfitting does not bias the slope estimates; in other words, the slope coefficients for a particular variable tend to remain accurate in an overfitted model. The main issue, however, is that standard errors become inefficient when a model is overfit. As a point of review, inefficient estimators often have larger variances than do efficient estimators. To understand the consequences of overfitting, it may be useful to consider the standard errors of the slopes from a multiple linear regression model. In chapter 3, we learned that the standard errors may be computed using the following equation:

$$\text{se}(\hat{\beta}_i) = \sqrt{\frac{\sum (y_i - \hat{y}_i)^2}{\sum (x_i - \bar{x})^2 \left(1 - \text{R}_i^2\right)(n - k - 1)}}$$

The quantity $(1 - \text{R}_i^2)$ is called the *tolerance* and is estimated from an auxiliary regression equation that is defined as $x_{1i} = \hat{\alpha} + \hat{\beta}_2 x_{2i} + \cdots \hat{\beta}_k x_{ki}$. Suppose that an extraneous variable (x_4) included in the model is strongly associated with an important variable (x_2) but is not associated with the outcome variable (y). Then, all else being equal, the tolerance for x_2 will be smaller (closer to zero) than if x_4 were not in the model and x_2's standard error in the linear regression model will be relatively larger. Figure 7.2 represents this situation.

The tolerance in the standard error computation for x_2 is represented by the overlap between the two explanatory variables. If the overlap is large, then the standard error can shift considerably, becoming large enough in some cases to affect the conclusions regarding the statistical significance of x_2's coefficient. (Keep in mind that we want substantial variation—that is, large circles—in the explanatory and outcome variables.) However, if there is not a statistical association between the explanatory variables—for example, corr(x_2, x_4) = 0—then the standard error associated with the important

Figure 7.2: Visualizing Tolerance

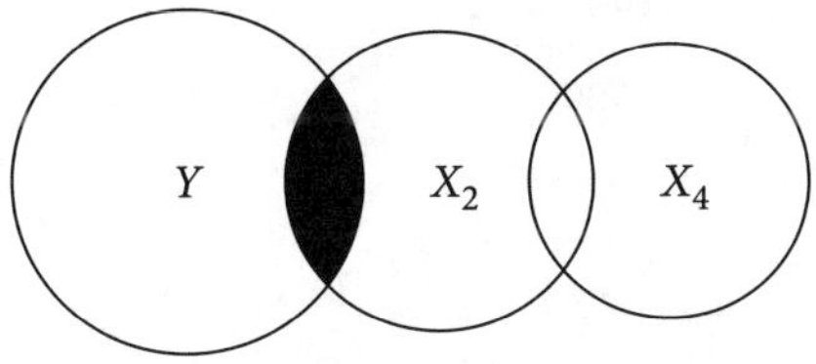

variable (x_2) is unaffected. In Figure 7.2, this would be presented by nonoverlapping circles between x_2 and x_4.

Underfitting—Or the Case of the Missing Variables

A more serious situation arises when we have left important variables out of the model—a problem often referred to as *omitted variable bias.* In this situation, we may reach the wrong conclusions about our ability to predict—or explain, for that matter—the outcome variable. This problem was highlighted in chapter 6 when we estimated the association between fundamentalist Christians and family income. There appeared to be a significant association between the two until we included education in the model. Then we learned that the association between fundamentalist Christians and family income is, in fact, partly spurious. Alternatively, we could say that education confounded the association between this limited measure of religious affiliation and income. There is an assumption of multiple linear regression models that applies here. In chapter 3, we learned the following: *the correlation between each explanatory variable and the error term—or residuals—is zero. If there is a correlation, we have left something important out of the model.*

Keep in mind that the error term in a regression model includes all of the factors, random or otherwise, that are associated with the outcome variable. Therefore, the information from omitted explanatory variables is included in the model. Suppose one of these omitted variables is associated with an explanatory variable we have put into the model. For instance, in the equation below, we have omitted x_3 and x_4. If x_3 is associated with y and with x_2, then the slope of x_2 will be biased. In general terms, we have violated a key assumption of the model. Unfortunately, we cannot know for certain the direction of the bias: Is the slope too large or too small? Is the true slope negative or positive? The answer to these questions depends on the joint association among x_2, x_3, and y.

$$y_i = \alpha + \beta_1 x_{1i} + \beta_2 x_{2i} + \varepsilon_i \qquad \text{where } \varepsilon_i = x_{3i} + x_{4i} + \text{random error}$$

Table 7.1: Bivariate Correlations

Variable	revhlth	educ	income06
educ	.2300		
	(.000)		
income06	.2064	.4044	
	(.000)	(.000)	
AfricanAm	−.0586	−.0965	−.234
	(.046)	(.000)	(.000)

Note: The numbers in parentheses are *p* values.

We saw an example of underfit in the last chapter, but let's explore another. The GSS 2010 data set includes a variable labeled *revhlth,* which is a personal assessment of how healthy an individual feels. Higher values on this variable indicate that the respondent feels healthier. There are studies that suggest that health is associated with education, income, racial/ethnic identity, and several other variables. We will examine some of these associations in a linear regression model. First, though, let's look at a bivariate correlation matrix of these variables—which can be found in Table 7.1.

The correlations suggest that income and education are positively associated with health, while being African American is negatively correlated with health. All three correlations are significant at $p < .05$ or better, as noted by the parenthetical numbers. The correlation between education and health and the correlation between income and health are similar, while the correlation between education and income, perhaps unsurprisingly, is fairly substantial ($r = 0.4044$). Unlike what we saw in the last chapter, we will begin with the full model and work backward.

The first thing to notice in the regression table (focusing on Model 2), once we account for education and income as predictors of health, is that the significant association from the correlation matrix between being African American and reporting negative health is not significant in the regression model in Table 7.2. However, both income and education have positive effects. In this situation, we say that the association between African American race/ethnicity and health is spurious; it is confounded by education and income.

However, suppose that, for some reason (perhaps we failed to read the literature on health carefully), we omit education and income from consideration. Perhaps we think that there is little health inequality or that insurance is the great equalizer for socioeconomic status, or we simply do not care much about education and income because, we argue, race/ethnicity is what really matters. Model 1, in Table 7.1, provides the regression results from a model that excludes education and income.

Table 7.2: Regression Model for Health, Models 1 and 2

Variable	Model 1	Model 2
African American	−0.200*	−0.096
	(0.096)	(0.095)
Education		0.060***
		(0.011)
Income		0.025***
		(0.006)
Intercept	3.637	2.341
N	1,068	1,068
R^2	.001	.069

$^{*}p < .05$. $^{***}p < .001$.

In this model, we find a statistically significant association between race and health. However, we know that this regression model is misspecified. We have a model that is underfit because it does not include variables that are associated not only with the outcome variable but also with race in such a way as to account for its presumed association with health. Not only is the slope for African American too large in Model 1 (−.200 → −.096) when we omit education and income from the model, but its standard error is slightly larger, too.

Another model to consider is one that omits African American race/ethnicity because of its nonsignificance in the presence of education and income (see Table 7.3). We have seen that this variable is significantly associated with life satisfaction whether we include income or not. In fact, its slope, standard error, t value, and p value are similar in both models, but does it affect the other variables?

We see that the results are not much different than when race/ethnicity is in the model. In Model 2, which has African American race/ethnicity in the model, the R^2 was .069, which reduced to .068 in Model 3, without race/ethnicity in the model. Furthermore, the slopes and standard errors associated with education and income are not affected much—changing by only .001 in either direction. Therefore, although this last model is also underfit, the consequences of specification error are less severe than when we leave out income. One lesson to learn from this exercise is that *specification error is a virtual certainty*. After all, we cannot hope or expect to include all variables that are associated with the outcome variable in the model. For example, very few social science data sets include genetic information, which has a strong role in predicting health outcomes. Thus, our goal is to strive for models with a low or manageable

Table 7.3: Regression Model for Health, Models 2 and 3

Variable	Model 2	Model 3
African American	−0.096	
	(0.095)	
Education	0.060***	0.059***
	(0.011)	(0.011)
Income	0.025***	0.026***
	(0.006)	(0.006)
Intercept	3.637	2.312
N	1,068	1,068
R^2	.069	.068

***$p < .001$.

amount of specification error and hope that we do not reach the wrong conclusions about the associations that the models do represent.

The next question to ask is whether the full model, which includes all three variables, is overfit. After all, African American race/ethnicity is not significantly associated with health once we consider education and income. On the one hand, it may not hurt our conclusions to include it in the model. On the other, it could make the other standard errors larger than they should be. At this point, the best way to check for overfitting is to estimate a linear regression model that omits the African American variable and look over the slopes and standard errors (see model 3 of Table 7.3). The results of such a model show that the slopes for education and income are virtually identical when race is omitted. Nonetheless, unless we were extreme in our desire to figure out the precise association between education, income, and this measure of health, it is a relative judgment call whether race/ethnicity needs to be in the model. However, it does little harm, and it is often better to err on the side of overfitting than underfitting (that is, omitted variable bias).

Endogeneity or Simultaneous Equations Bias

Recall that, at the beginning of chapter 2, we mentioned that another name for the outcome variable is the *endogenous* variable. The prefix *endo-* refers to something that is inside or within. Thus, when referring to an endogenous variable or outcome, we mean that it is produced within the system or within the equation. Our linear regression equation, you will have noticed by now, assumes that the outcome variable, but not the explanatory variables, is "produced" within the system. The explanatory variables are considered *exogenous*, or produced by factors outside of the model or system. We may also say that the explanatory variables are *predetermined*—"*pre*" meaning that they are formed outside the system implied by the model.

However, think about a set of variables such as education, income, race/ethnicity, and health. Models are often proposed that define these variables, or many other sets of variables, as producing one another in some way. For example, we could revise our earlier model to include gender:

$$\text{Health}_i = \alpha + \beta_1(income_i) + \beta_2(gender_i) + \varepsilon_i$$

This model implies that income and gender independently combine to affect health, but suppose that income and gender are not independent; rather, because of various historical and social factors that are associated with gender in the United States and elsewhere, income is, in part, a product of an individual's gender. In this situation, we say that income is endogenous in the system specified by the equation because it may be predicted by gender. The problem for the linear regression model is that the estimated slope for health in the above equation will then be biased. A second potential source of endogeneity involves whether we have specified the correct ordering of variables. Does the outcome variable truly "depend" on the explanatory variables? Or could one or more of the explanatory variables "depend" on the presumed outcome variable? For instance, suppose we wish to estimate a model with adolescent drug use as the outcome variable and friends' drug use as the explanatory variable. We may assume that one's friends influence one's behavior to a certain degree, thus leading to this model specification, but it may also be true that one's choice of friends depends on one's behavior. Hence, those who use drugs are more likely to choose friends who also use drugs. This issue implies that drug use and friends' drug use are involved in what is known as *reciprocal causation*.

We won't pursue the issue of endogeneity in any more detail at this point, because it involves both a thorny theoretical issue and a complex statistical issue. Theoretically, we need to think carefully about the models we wish to estimate and consider whether one or more of the explanatory variables is potentially endogenous. When we estimate models with variables such as education, income, and life satisfaction, or one's own behavior and one's friends' behavior, we should think about the order and intrasystem relationships among the variables. Are the explanatory variables truly independent? Are they determined outside the model? Or could one or more of the variables "depend" on one or more of the other explanatory variables? Moreover, could variables outside the system affect the explanatory variables? The answer is almost always yes, so correct model specification and lack of endogeneity require a properly fit model.

Endogeneity is a complex statistical issue because it requires complicated systems of equations to specify correctly. We will learn in chapter 8 about two statistical models, instrument variables with two-stage least squares and structural equation models, that are designed to address endogeneity in regression models. Statistical programs such as Stata and SPSS typically have other approaches for addressing endogeneity—although they are outside the scope of this text.

How Do We Assess Specification Error, and What Do We Do about It?

Clearly, specification error is a problem we wish to avoid, but as implied earlier, considering all its sources is not an easy task. The best advice is to always strive for a clear and convincing conceptual model or theory to guide the analysis. Conceptual models are the result of previous studies and reasoned thought about the processes involved in creating an association between two or more variables. Therefore, you should not even begin to estimate regression models until you have carefully read the research literature that addresses the outcome variable. It is almost always a good idea to read the literature on your key explanatory variables as well. You may find some unexpected variables that are associated with the explanatory variables you plan to use in your regression model.

Unfortunately, we cannot even hope to include all the variables that might be associated with the explanatory and outcome variables; as mentioned earlier, specification error is always lurking—we just hope to minimize it. There are, however, some tools that are handy for assessing whether specification error is affecting the results of the regression model. The first such tool may be understood by considering, as an example, the equation shown earlier in the chapter:

$$y_i = \alpha + \beta_1 x_{1i} + \beta_2 x_{2i} + \varepsilon_i \qquad \text{where } \varepsilon_i = x_{3i} + x_{4i} + \text{random error}$$

When we initially saw this equation, we were interested in what happens to the regression model when x_2 and x_4 are correlated. Although a correlation might affect the conclusions, how might we test whether x_4 has an influence if we do not measure it directly? Can we figure out a way to measure the error terms (residuals) so that we may, at least indirectly, examine the possible association between x_2 and x_4? As we learned in earlier chapters, we may use the predicted values from the model to compute the residuals:

$$residual_i(\varepsilon_i) = (y_i - \hat{y}_i)$$

In chapter 3, we saw that Stata allows us to predict the residuals from a linear regression model by using the `predict` postcommand. (Using the `rstandard` or `rstudent`

Figure 7.3: Stata Output for the linktest Command

. linktest

Source	SS	df	MS		
				Number of obs =	1068
				F (2, 1065) =	40.58
Model	82.829088	2	41.414544	Prob > F =	0.0000
Residual	1086.91155	1065	1.02057422	R-squared =	0.0708
				Adj R-squared =	0.0691
Total	1169.74064	1067	1.09628926	Root MSE =	1.0102

revhlth	Coef.	Std. Err.	t	P>|t|	[95% Conf.	Interval]
_hat	−2.10088	2.132868	−0.99	0.325	−6.285981	2.084221
_hatsq	.4355005	.299132	1.46	0.146	−.1514546	1.022456
_cons	5.485735	3.789944	1.45	0.148	−1.95087	12.92234

options, you may also calculate *standardized* residuals, which are the residuals transformed into *z* scores; or *studentized* residuals, which are the residuals transformed into Student's *t* scores.) SPSS allows us to save the residuals from a linear regression model using *Save-Predicted Values* in the Linear Regression drop-down menu and then checking the *Residuals-Unstandardized* box. (You may also save the *standardized* residuals, which are residuals transformed into *z* scores; or, the *studentized* residual, which are transformed into Student's *t* scores.) We may then assess whether the explanatory variables included in our model are correlated with the residuals. Unfortunately, because of the way the ordinary least squares estimators are derived, there is rarely, if ever, a linear association between the *x*s and the residuals. Sometimes, though, we might find a nonlinear association. We will learn more about examining potential nonlinear associations in chapter 10.

Stata has several regression postcommands that may be used to test for specification errors. The most straightforward, and probably the most useful, test is Stata's `linktest` command. This command is based on the notion that if a regression is properly specified, we should not be able to find any additional explanatory variables that are significant except by chance. The test creates two new variables, the variable of prediction, *_hat*, and the variable of squared prediction, *_hatsq*. The model is then reestimated using these two variables as predictors. The first, *_hat*, should be significant because it is the predicted values. Conversely, *_hatsq* shouldn't be significant, because if our model is specified correctly, the squared predictions should not have any explanatory power. That is, we would not expect *_hatsq* to be a significant predictor if our model is specified correctly. Therefore, we should look at the *p* value for *_hatsq*.

Figure 7.3 shows what `linktest` provides after our fully specified health regression model with *educ*, *income06*, and *AfricanAm* as explanatory variables. The results

Figure 7.4: Stata Output for the Omitted Variable Test

```
. ovtest
Ramsey RESET test using powers of the fitted values of revhlth
    Ho: model has no omitted variables
            F(3, 1061)  =      1.66
              Prob > F  =      0.1730
```

suggest that there is not an omitted variable or some other specification problem because the *_hatsq* coefficient has a *p* value of .146. An alternative but similar approach is the RESET (*regression specification error test*) option, which is implemented in Stata using the `ovtest` postregression command (as shown in Figure 7.4).

The null hypothesis for this test is that there is no specification error. It is not rejected in this situation, thus reaffirming the finding that there is not specification error in this model. A third type of test that we won't discuss here, but that is available in Stata, is called a *Hausman test*. It is more generally applicable than the other two tests. Unfortunately, these tests do not tell us whether the misspecification bias is due to omitted variables or whether it exists because we have the wrong functional form for one or more of the variables in the model. To test for problems with functional forms, we may be better off examining partial residual plots and the distribution of the residuals (see chapter 10). Theory remains the most important diagnostic tool for misspecification bias, though.

Unfortunately, testing for specification error in SPSS is much more difficult than in Stata. For example, the Hausman test requires a substantial amount of programming to get from the program. Further, conducting these tests in SPSS requires several assumptions about your variables that are unlikely to hold. In chapter 8, we will introduce a different kind of test, the *Durbin-Wu-Hausman test*, which is more feasible in SPSS.

Thankfully, testing for underfit or overfit is much easier when you have access to all (or most) of the important explanatory variables in the data set. We obviously know that a regression model is underfit if we add a variable that is significantly associated with the outcome variable (a *t* test or nested *F* test may be used to confirm this; see chapter 5). Moreover, some researchers recommend using the adjusted R^2 to compare models. This should be used only when one of the models is nested within the other. If we add, say, a set of explanatory variables and find that the adjusted R^2 increases, then we may conclude that the nested model is underfit. If, on the other hand, we take away a set of variables (perhaps because their associated *p* values are large) and find that the adjusted R^2 does not decrease, then we may conclude that the model is overfit.

However, it is not a good idea to rely exclusively on the adjusted R^2. It is better to combine an evaluation of this statistic with a nested *F* test and a careful review of the slopes and standard errors of the remaining explanatory variables before concluding that a particular regression model is overfit or underfit. However, we must also beware of overdoing it, of testing so many models by adding and subtracting various sets of explanatory variables that we cannot decide on the appropriate model.

Variable Selection Procedures

The aim of minimizing underfit and overfit is related to an important topic in regression modeling: variable selection procedures. Unfortunately, many, if not most, statisticians' guides to variable selection are so automated as to violate a point made earlier in this chapter: Let your theory or conceptual model guide the selection of variables. Although we do not recommend the following procedures, they are discussed so often that they are worth a brief review.

The first procedure is known as *forward selection*. In this approach, explanatory variables are added to the model based on their correlations with the unexplained component of the outcome variable. Yet this relies on biased estimates at each point in the selection process, so it should be avoided. A type of forward selection that you might encounter is *stepwise regression*, wherein partial *F* statistics are computed as variables are removed and put back in the model. For various reasons that we discuss later, this should also be avoided. The second type is known as *backward selection*. In this approach, the analyst begins with all the explanatory variables in a model and then selectively removes those with slopes that are not statistically significant. This selection procedure is reasonable if you wish to have the best predictive model, but it is overly automated and takes the important conceptual work out of the hands of the researcher. Nonetheless, we will consider some steps that help make backward selection somewhat better than more arbitrary or biased approaches.

There are several steps to selecting variables if we wish to rely on backward selection: First, as mentioned earlier, estimate a regression model with all the explanatory variables. Second, choose a fit measure to compare models. Third, use backward selection to remove variables—this may be done one at a time or in chunks—from the model. Fourth, assess the reliability of the preferred model with a split sample procedure. This last step often is reasonable if we want to verify the reliability of a regression model regardless of how we decided which variables to include. If the software allows it (Stata does so using the `sample` command; SPSS permits this using the drop-down menu *Data-Select Cases-Random sample of cases*), select a random subsample of observations from the data set (normally 50 percent) at the variable selection stage and then, after choosing the best model, estimate it by using the remaining subsample of observations. If the models are the same (or quite similar), then we may have confidence that we have estimated the best-fitting model.

These steps are fairly simple when using statistical software, but we should say a few more words about the second step: choosing a fit measure to compare models. We have seen a couple of these measures already: the adjusted R^2 and nested F tests. If we compare the full model to its nested models by using these approaches, we can guard against including extraneous variables in the model. Moreover, some statisticians recommend comparing the mean square error (MSE) or its square root (S_E) (see chapter 4):

$$\text{MSE} = \frac{SSE}{n-k-1} \qquad \sqrt{\text{MSE}} = S_E$$

Looking for the smallest MSE or S_E from a series of nested models is a reasonable step, given that, assuming we wish to make good predictions and find strong statistical associations, we would like to have the least amount of variation among the residuals.

There is another common measure that is related to these other measures: Mallow's C_p. An advantage of Mallow's C_p is that it is minimized when the model has only statistically significant variables. Mallow's C_p is computed as:

$$C_p = \frac{\text{SSE}(p)}{\text{MSE}(k)} - [n-2(p+1)]$$

The SSE(p) is from the restricted model, and the MSE(k) is from the full model. Using this measure does not require the model with all the explanatory variables and some model nested in it. It may include any two models that use subsets of the explanatory variables as long as at least one is nested within another. Nonetheless, Mallow's C_p is used most often to compare the complete or full model with various models nested within it.

Here is an example of three models that use three of these model selection fit statistics: the adjusted R^2, S_E, and Mallow's C_p. The models use the GSS 2010 data set and are designed to predict number of poor physical health days (*physhlth*). Suppose that, on the basis of previous research, we think that the following explanatory variables are useful predictors of poor physical health: income, education, race/ethnicity, and gender. However, someone convinces us that we should use a backward selection procedure to test several combinations of these variables.

Both Stata and SPSS offer procedures that can test various combinations of variables. In Stata, if we estimate the model with all four explanatory variables using the stepwise command, we get a "best-fitting" model:

```
stepwise, pr(0.2): regress physhlth AfricanAm othrace gender
income06 educ
```

The pr subcommand requests Stata's backward selection model. The 0.2 tells Stata to retain a variable only if its coefficient has a p value of no larger than .2. We may

Figure 7.5: Stepwise Regression Output in Stata

```
. stepwise, pr(0.2): regress physhlth AfricanAm othrace gender income06 educ
                    begin with full model
p = 0.2432 >= 0.2000 removing AfricanAm
```

Source	SS	df	MS
Model	783.502511	4	195.875628
Residual	43734.7056	1057	41.3762588
Total	44518.2081	1061	41.9587258

Number of obs	=	1062
F (4, 1057)	=	4.73
Prob > F	=	0.0009
R-squared	=	0.0176
Adj R-squared	=	0.0139
Root MSE	=	6.4324

physhlth	Coef.	Std. Err.	t	P>|t|	[95% Conf.	Interval]
educ	−.1082827	.0713625	−1.52	0.129	−.2483109	.0317456
othrace	.9832991	.6716004	1.46	0.143	−.3345225	2.301121
gender	−1.076731	.401805	−2.68	0.007	−1.865158	−.2883052
income06	−.0700382	.0405347	−1.73	0.084	−.1495758	.0094994
_cons	6.190929	.9996068	6.19	0.000	4.22949	8.152368

also use the `pe` subcommand to request forward selection. See Stata's help menu (help `stepwise`) for additional options. Using backward selection with stepwise regression returns a model with all the explanatory variables included except for being African American, as shown in Figure 7.5.

In SPSS, we can use the same procedure by running the regression with *AfricanAm, othrace, gender, income06,* and *educ* as predictors of *physhlth.* Then, select Method: Backward (the default is Enter) in the middle of the window, click *OK,* and check the SPSS output window. The output should look like that shown in Figure 7.6.

The lesson from using these automated procedures is that some of them, such as backward selection, are reasonable tools if your goal is to come up with a model that includes the set of explanatory variables that offers the most predictive power. However, these procedures cannot substitute for a good theoretical or conceptual model that not only predicts the outcome variable but, more important, *explains* why it is associated with the explanatory variables. Moreover, they may provide misleading results because they are designed to fit the sample data, thus overstating how precise the results appear to be when we wish to make inferences to the population from which the sample was drawn (for more information, see Fox, 1997).

There are other reasons to avoid stepwise selection procedures:

- The model's R^2 values are biased.
- The coefficients' confidence intervals are biased (they are too narrow).
- The p values are not meaningful without complicated corrections.
- They have serious problems when there is collinearity (see chapter 9).

Figure 7.6: Stepwise Regression Output in SPSS

Model Summary

Model	R	R Squared	Adjusted R Squared	Std. Error of the Estimate
1	.137[a]	.019	.014	6.43133
2	.133[b]	.018	.014	6.43244
3	.125[c]	.016	.013	6.43591
4	.115[d]	.013	.011	6.44034

ANOVA

Model		Sum of Squares	df	Mean Square	F	Sig.
1	Regression	839.890	5	167.978	4.061	.001[a]
	Residual	43678.318	1056	41.362		
	Total	44518.208	1061			
2	Regression	783.503	4	195.876	4.734	.001[b]
	Residual	43734.706	1057	41.376		
	Total	44518.208	1061			
3	Regression	694.807	3	231.602	5.591	.001[c]
	Residual	43823.401	1058	41.421		
	Total	44518.208	1061			
4	Regression	592.986	2	296.493	7.148	.001[d]
	Residual	43925.222	1059	41.478		
	Total	44518.208	1061			

[a]Predictors: (Constant), total family income, RECODE of hhrace (race of household), RECODE of sex (respondents sex), RECODE of hhrace (race of husehold), highest year of school complete.

[b]Predictors: (Constant), total family income, RECODE of hhrace (race of household), RECODE of sex (respondents sex), highest year of school complete.

[c]Predictors: (Constant), total family income, RECODE of sex (respondents sex), highest year of school complete.

[d]Predictors: (Constant), total family income, RECODE of sex (respondents sex).

- The methods used to compare models (for example, nested *F* tests) are supposed to be used only with prespecified models.
- They make us lazy and overconfident: We need to think carefully about the conceptual model and not rely on the software to judge what is and is not appropriate.

Key Terms

The following is a list of key terms and concepts in chapter 7. You can find definitions to these words throughout the chapter.

- Backward selection
- Curvilinear relationship
- Endogeneity
- Exogenous
- Forward selection
- Omitted variable bias
- Overfitting
- Reciprocal causation
- Specification error
- Stepwise regression
- Tolerance
- Underfitting

Analysis Exercises

1. Estimate a model with occupational prestige (*prestg80*) as the outcome variable and the race/ethnicity dummy variables (*AfricanAm*, *othrace*), gender (*gender*), and mental health (*mntlhlth*) as explanatory variables. In an additional model, include education (*educ*) as a control variable. What do the results tell you about the possibility of underfit?
2. What does the model estimated in question 1 potentially tell us about the endogeneity of occupational prestige?

Solutions

1. The inclusion of education into the model renders mental health nonsignificant, although it was significant when education was not included in the model. This suggests that we omitted an important variable and that our first model (which did not include education) was, in fact, underfit.
2. One possibility is that some sort of reciprocal causation is present in the model. Is it possible that occupational prestige could lead to poorer mental health? For example, could people in very high- or very low-prestige jobs be more likely to have poor mental health than others? A second issue might be that educational attainment is associated with race and gender in the United States, suggesting the possibility that occupational prestige is an endogenous process.

CHAPTER 8 MEASUREMENT ERRORS IN LINEAR REGRESSION MODELS

Chapter Objectives

In this chapter, we discuss the many issues that can arise from mismeasured explanatory and outcome variables. We also consider some ways to overcome this problem.

In chapter 8, we will

- discuss measurement error in social work and the social sciences
- cover the problems created by mismeasured explanatory and outcome variables
- introduce latent variables
- illustrate the usefulness of the two-stage least squares regression.

How do we measure concepts in the social and behavioral sciences? There are numerous things we might wish to measure, which either lack a clear definition or are ambiguous in how they would be measured. Concepts such as self-esteem, depression, happiness, social capital, antisocial behavior, aggression, impulsivity, political ideology, marital satisfaction, social support, and dozens of other phenomena do not easily lend themselves to measurement by social scientists. We do a much better job of measuring concepts such as monthly income in dollars, official rates of crime and suicide, or years of education, as these are more tangible and there are widely accepted ways of assessing the quantity of each. However, even these are often measured poorly because we must rely on people to report them accurately. Sometimes people do not report these things with care, or there are sources of error that are out of their control.

Researchers often measure various concepts by using self-report instruments. To measure adolescent drug use, for example, they usually ask a sample of adolescents to report whether and how often they've used alcohol, cigarettes, marijuana, cocaine, or other illicit drugs. To measure happiness, researchers ask people whether they are happy in particular spheres of their lives (for example, job, family), or they ask a global question such as, "Do you consider yourself a happy person?" There are probably many ways that happiness or some similar concept such as life satisfaction can be measured, but generally we must be content with asking people about themselves. Even standardized assessment instruments, like a depression scale, rely on self-reporting. In fact, only rarely is there an external instrument that is useful for verifying what these people tell us. An interesting exception involves recent research on illicit drug use. There are now widely available saliva and hair tests used to determine whether people have recently used illegal substances such as marijuana or cocaine. Of course, we could hook respondents up to lie detector machines—which are actually quite accurate—and then ask questions. This would likely deter most people from participating in surveys, however.

The literature on measurement in the social and behavioral sciences is huge, and we cannot even begin to cover all the important topics. Our concern is with a narrow statistical issue—but one of utmost importance—that involves measurement. Known in general terms as *measurement error* or *errors in variables*, we are interested in learning about what happens to the regression model when the instruments used to measure the variables are not accurate (see Beider, Groves, Lyberg, Mathiowetz, & Sudman, 2004). There are various sources of measurement error, including *method error*, which occurs because we simply do not have accurate measuring instruments (such as scales or rulers); and *trait error*, which occurs because people sometimes misunderstand the questions that researchers ask or because respondents don't answer them accurately for some reason (lack of interest or attention, exhaustion from answering lots of questions). Sometimes people do not wish to tell researchers the truth or provide information about personal topics. One of the most personal topics, it seems, is personal income. If you ask questions about a person's income, be prepared for many refusals or downright deceptions.

Another source of measurement error is recording or coding errors, such as when a data entry person forgets to hit the decimal key, so an income of $1,900.32 per month becomes an income of $190,032 per month. Normally, this type of measurement error is easy to detect during the data screening phase. If we make it a point to always check the distributions of all the variables in our analysis using means, medians, standard deviations, ranges, box-and-whisker plots, and the other exploratory techniques and are familiar with the instruments used to measure the variables, we will usually catch coding errors. (Don't forget to always check the missing data codes in your data set!)

Suppose that we have carefully screened our data for coding errors and fixed all the apparent problems. We may still suspect that there is measurement error in the variables for all the other reasons we have discussed so far. Is there anything we can do about it? We will discuss two situations that lead to distinct solutions: (1) The outcome variable, y, is measured with error; (2) one or more of the explanatory variables, x, is measured with error.

Before discussing these two situations, let's revisit some of the assumptions (see chapter 3) of linear regression analysis that relate to measurement error. First, there is an assumption that is almost never followed in the social sciences that the x variables are fixed (the technical term is *nonstochastic*). This means that researchers have control over the x variables. For example, in agricultural studies researchers may try out different values of the explanatory variable (such as by applying different amounts of fertilizer and water to plants and examining their growth) and determine, with quite a bit of accuracy, the association between x and y. Second, we assume that the x and y variables are measured without error. We will learn in this chapter what to do when this assumption is violated. Third, x is uncorrelated with the error term. We saw one type of violation of this assumption in chapter 7 when addressing specification error.

The first assumption is not typically met in the social sciences. There is certainly experimental research—especially in psychology but also in social work, sociology, political science, and economics—that manipulates explanatory variables, but most social science research relies on random samples to consider a variety of values of explanatory variables. There are important ethical limits faced by social and behavioral researchers. For example, we should not—perhaps cannot—manipulate an adolescent's friends in studies of peers and delinquency. Another limitation involves the fact that there are many interesting variables that cannot be manipulated. For instance, we cannot control or manipulate a person's race or gender. Although deliberate manipulation of the xs is preferred, statistical control is often the best substitute.

Because the third assumption was the main focus of chapter 7, much of our discussion in this chapter centers on the second assumption. We will learn that measurement error can cause several problems for linear regression models. As with specification error, you should realize that we always have measurement error, either because our concepts are not well defined or for some systematic reason (for example, our instruments are not accurate enough). It is important that we seek to reduce it to a manageable level.

The Outcome Variable Is Measured with Error

Suppose there is an outcome variable, such as happiness, that, in all likelihood, is measured with error. For now, we will assume that our explanatory variables, say education and income, are measured without error. Furthermore, we will assume that somewhere within the conceptual measurement of happiness is a true measure, or what is often called the *true score*. Let's represent the outcome variable as the sum of two components, true score plus error:

$$y_i^* = y\,(\text{truescore})_i + u\,(\text{error})_i$$

Using these terms, what does the linear regression model look like? Here is one way to represent it:

$$y_i^* = \alpha + \beta_1 x_{1i} + \beta_2 x_{2i} + \varepsilon_i = y_i + u_i = \alpha + \beta_1 x_{1i} + \beta_2 x_{2i} + \varepsilon_i$$

Now, let's use basic algebra to subtract u (the error) from both sides of the equation. By solving the equation, we obtain

$$y_i = \alpha + \beta_1 x_{1i} + \beta_2 x_{2i} + \varepsilon_i + (-u_i)$$

Therefore, the error in measuring y can be considered just another component of the error term, or simply another source of error in our model. As long as this source of error is not associated systematically with the explanatory variables—that is, corr(x_{1i}, u_i) = 0; corr(x_{2i}, u_i) = 0—we have not violated a key assumption of the model, and it is unlikely that the slope estimates will be biased. However, if the xs are correlated with the measurement error in y, we will have introduced specification error (as described in chapter 7). Unfortunately, there is rarely any way to include this error explicitly in the regression model to solve the specification problem.

However, even when there is no correlation between the xs and the error in y, the R^2 tends to be lower than if measurement error is not present. In the presence of error in y, the SSE is larger because there is more variation in the residuals that is not accounted for by the model. Moreover, the standard errors of the slope estimates are inflated, thus making it harder to find statistically significant slope coefficients. Recall that the standard error formula is

$$\text{se}(\hat{\beta}_i) = \sqrt{\frac{\sum (y_i - \hat{y}_i)^2}{\sum (x_i - \bar{x})^2 (1 - \text{R}_i^2)(n - k - 1)}}$$

Notice what happens when the SSE $\{\Sigma(y_i - \hat{y}_i)^2\}$ is larger. All else being equal, the standard error is also larger. All is not lost, though. If you obtain statistically significant slope coefficients, you then have a measure of confidence that you would find significant results even if there was no measurement error in y.

The Explanatory Variables Are Measured with Error

More problematic is when one or more of the explanatory variables is measured with error. For now, let's assume that the outcome variable is measured without error, so we have only y(true score) to deal with. However, x, as we observe it, is made up of a true score plus error $\{x_{1i}^* = x_{1i} + v_i\}$.

$$y_i = \alpha + \beta_1 x_{1i}^* + \varepsilon_i = y_i = \alpha + \beta_1 (x_{1i} + v_{1i}) + \varepsilon_i$$

Distributing by the slope, we have the following regression equation:

$$y_i = \alpha + \beta_1 x_{1i} + \beta_1 v_{1i} + \varepsilon_{ii}$$

The error term now has two components ($\beta_1 v_{1i}$ and ε_i), with x correlated with at least one of them. In this situation, the estimated slope is biased, usually toward zero. The degree of bias is typically unknown, though; it depends on how much error there is in the measurement of x. The standard error of the slope is also biased. Usually, the t value is smaller (because of a larger standard error) than it would be if there was no measurement error in x. Combining these two problems—smaller slopes and larger standard errors—shows that statistically significant slopes are difficult to find when x is measured with error. This does not mean, however, that there is not a true statistically significant association between x and y; only that you may not be able to detect it if there is sufficient error in x to bias the slope and standard error.

Let's suppose that we have a multiple linear regression model and that several of the explanatory variables are measured with error. It should be apparent that this compounds the problem, sometimes tremendously, but what happens when both the outcome variable and several of the explanatory variables are measured with error? Now we have a truly bad situation: The R^2 is biased, the slopes are biased, and the standard errors are biased. We can't trust much about such a model. Yet it is not uncommon in the social and behavioral sciences to have measurement error in all of the variables. A major problem is that we don't often know if and how much measurement error we have. Fortunately, there are several techniques designed to address this problem.

What to Do about Measurement Error

We will discuss five (partial) solutions to measurement error. The first one is easy to state yet slippery to implement: Have good measuring instruments. This is not meant as a flippant response to an important question. The development and testing of instruments for use in surveys and social and behavioral experiments is a huge area of research. It is mostly within the domain of psychology and cognitive studies but also draws from a variety of fields, such as social work, sociology, marketing, political science, and opinion research (see Krosnick & Fabrigar, 2005). We have neither the space nor the ability in this brief chapter to go over the many aspects of good measurement, including various topics such as the validity and reliability of instruments. One piece of advice, though, is that it is better to use a well-established instrument to measure a phenomenon than to try to make one up for your study. For instance, if we wish to measure symptoms of depression, using an instrument such as the Hamilton Depression Inventory is better than trying to develop a brand new scale for a study.

The next two approaches are not normally very useful, but you may run into situations where knowing about them comes in handy. The first is known as *inverse least squares* regression. Its logic is pretty straightforward. Suppose you wish to estimate a regression model with only one explanatory variable, but you know that this variable is measured with error. Fortunately, the outcome variable is not measured with error (an admittedly rare situation). In this situation, we may switch the explanatory and outcome variables and estimate a regression model:

$$x_i = \alpha + \beta_1 y_i$$

However, is this the slope you wish to use? Probably not. Rather, recalling elementary algebra, we may reorder the equation (after estimating the slope) so that the x appears on the right-hand side and the y appears on the left-hand side. In this manipulation, the slope for x becomes $1/\beta_1$ and it is not biased (although the standard error is not accurate). Note also that the intercept is not correct in this type of least squares regression.

The second method assumes that we have some external knowledge that is not normally available. First, though, we need to learn about a term in measurement theory known as *reliability*. In general usage, *reliability* refers to the reproducibility of a measuring instrument. The most common type is *test–retest reliability*, which is a quantitative approach to determining whether an instrument measures the same thing at time 1 that it measures at time 2. Assuming a continuous variable, researchers usually use the Pearson's correlation coefficient of the two measurement points to determine the reliability.

A common view of reliability is that it measures the ratio of the true score to the measured score. Using the notation introduced earlier (with the subscripts omitted), this is written as

$$\mathrm{r}_{xx^*} = \frac{x(\text{true score})}{x(\text{true score}) + v(\text{error})} = \frac{\mathrm{s}^2 x}{\mathrm{s}^2 x^*}$$

In this equation, r represents the reliability, not the Pearson's correlation coefficient (although there are statistical similarities). The s^2 in the numerator and denominator are variance measures. It should be clear, and this agrees with what we have been claiming throughout this chapter, that reliability increases as the error in measurement decreases.

Although rare, we occasionally may have information on the reliability from other studies. Suppose, for example, that we are conducting a study that uses Mudd's Scale of Joy to measure happiness among teenage boys and girls (this is not an actual scale). Several very sophisticated studies have found that when Mudd's Scale is used among this population, it yields a reliability coefficient (r_{xx^*}) of 0.8. (We won't bother trying to figure out how these other researchers came up with this reliability coefficient.) Because errors in measuring the *x* variable tend to bias the slope downward, our observed slope will be too small if *x* has measurement error, but we now know the degree of bias because of the reliability coefficient. Therefore, we may multiply the observed slope by the inverse of the reliability coefficient to obtain the unbiased slope. Assume that our model suggests that high levels of joy reported by adolescents predict high grade point average. We estimate a linear regression model and obtain an unstandardized slope of 1.45. Multiplying this slope by $1/0.8 = 1.25$ yields the unbiased slope of $\{1.45 \times 1.25\} = 1.81$. Unfortunately, this approach becomes unwieldy to use when there is more than one explanatory variable in a model.

Latent Variables as a Solution to Measurement Error

The term *latent* means hidden, concealed, or present but not revealed to the senses. It has been adopted in applied statistical analysis to refer to variables that exist in some indistinct sense but are not detectable using observable methods. In the social and behavioral sciences, one might argue, virtually all variables are latent: depression, anxiety, happiness, attitudes, antisocial behavior, prosocial behavior, and so on. We do not see, hear, smell, taste, or touch these variables. Perhaps we can be confident in our ability to recognize phenomena, like Supreme Court Justice Potter Stewart when he famously wrote that he could detect pornography in motion pictures because, "I know it when I see it." Most of us, though, would likely have a difficult time coming up with a universal way to put fixed boundaries around many of the phenomena of interest in the social and behavioral sciences. One solution to this slippery problem is to measure concepts indirectly through the use of variables that we can measure well. Measurement of concepts using latent variables provides the most common solution.

As the name implies, latent variables are thought to exist but are hidden from our senses (actually, we assume they are hidden from the direct capabilities of our measuring instruments). The accepted logic is that we can measure aspects of some latent concept. By accumulating information from these measurable aspects, we may construct a latent variable. The statistical area that permits this approach is known as *multivariate analysis* because it is concerned with variance that is shared among multiple variables. Under the broad area of multivariate analysis fall statistical techniques such as principal component analysis, factor analysis, latent variable analysis, log-linear analysis, and several other statistical procedures.

Figure 8.1: Visualizing Shared Variance

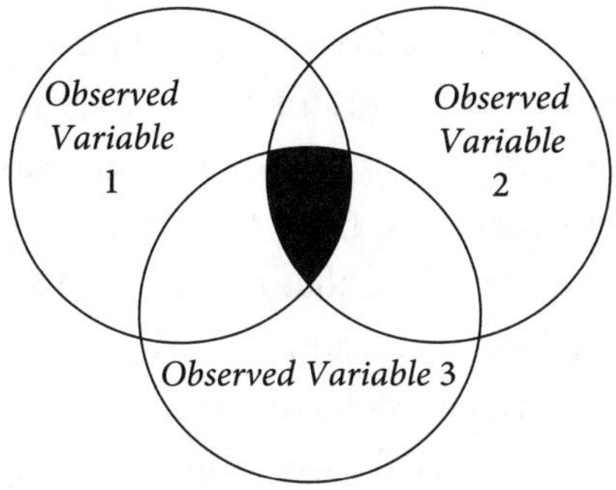

Figure 8.2: Latent Variables

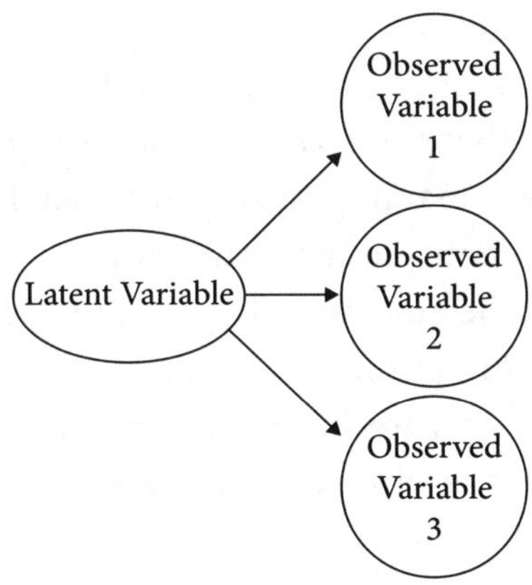

We will discuss very briefly a frequently used technique known as *factor analysis*. Factor analysis is designed to take *observed* or *manifest* variables and reduce them to a set of latent variables that underlie the observed variables. There are many books that focus specifically on factor analysis (for more information, see Bartholomew, 1987; Kline, 1999; Thompson, 2004). Perhaps the simplest way of understanding this statistical technique is to consider that variables that are similar share variance. If there is sufficient shared variance among a set of observed variables, we claim that this shared variance represents the latent variable. We assume that, conceptually speaking, the latent variable predicts the set of observed variables. Figures 8.1 and 8.2 represent two simplified models of a latent variable. Figure 8.1 shows the overlap of three observed variables; the shaded area of overlap represents the latent variable, or the area of shared variability. Figure 8.2 shows how latent variables are often represented in books and articles. Notice the direction of the arrows. They imply that the latent variable (recall that it is not directly observed) predicts the observed variables.

Both SPSS and Stata provide several factor analysis techniques in their packages. However, when used to measure latent variables, many researchers rely on specialized software such as AMOS, LISREL, COSAN, EQS, Latent Gold, or Mplus to conduct the analysis. An advantage of these programs is that they will simultaneously estimate factor analyses to estimate the latent variables and regression models to test the associations among a set of latent variables. For example, if we used 10 observed variables to measure aspects of anxiety, assuming that a latent variable—or the true measure of anxiety—explains their shared variance, and 12 variables to measure self-esteem, then a latent variable analysis program allows us to test the measurement properties of these two latent variables and their statistical association by using a regression model. One of the advantages of this approach—which is often termed *structural equation modeling* (SEM)—is that it may be used to estimate more than one regression equation at a time, or what are called *simultaneous equations*. This provides one solution to the endogeneity problem discussed in chapter 7.

However, you may be asking; What happened to the measurement error in the latent variable? The idea is that the shared variance represents the true score of a latent variable, whereas the remaining unshared variability represents the error. In Figure 8.1, the shaded area where all three variables overlap represents the true score for the latent variable and the nonshaded areas represent measurement error. Of course, we must make the assumption that the shared variance truly represents something that is significant and that we have correctly labeled it. We also assume that the measurement errors for the observed variables are independent, although this assumption may be relaxed in SEM.

The second latent variable approach has been popular in econometrics. In this approach, the latent variables are based on *instrumental variables*. An example is the easiest way to describe instrumental variables. Suppose we have an explanatory variable that is measured with error. We know that if we place it in a regression model, the slopes and standard errors will be biased. However, we think we have found another set of variables that have the following properties: (1) They are highly correlated with the error-contaminated explanatory variable; and (2) they are uncorrelated with the error term (which, you'll recall, represents all the other influences on y that are not in the model). If we can find such a set of variables, we may use them in an ordinary least squares (OLS) regression model to predict the explanatory variable, use the predicted values to estimate a latent variable as a substitute for the observed explanatory variable, and use this latent variable in an OLS regression model to predict y. One common estimation technique that uses instrumental variables is known as *two-stage least squares* because it uses the two OLS regression stages just described. We assume that the latent variable is a linear combination of the instruments; thus, we can estimate the association with a linear model.

Figure 8.3: Correlation Matrix Produced by Stata

	revhlth	prestg80	educ	income06
revhlth	1.0000			
prestg80	0.1850	1.0000		
educ	0.2279	0.5203	1.0000	
income06	0.1971	0.4286	0.4136	1.0000

Here are the steps in a more systematic fashion. We will use the letter z to represent the instrumental variables and x to present the error-contaminated explanatory variable. The steps in two-stage least squares are as follows (the i subscripts are omitted):

1. Estimate $x_1 = \alpha_\mathrm{p} + \beta_{\mathrm{p1}} z_1 + \beta_{\mathrm{p2}} z_2 + \cdots + \beta_{\mathrm{pq}} z_\mathrm{q}$.
2. Save the predicted values from this OLS regression model.
3. Using OLS, estimate $y = \alpha_\mathrm{r} + \beta_{\mathrm{r1}} x'_1 + \beta_{\mathrm{r2}} x_2 + \cdots + \beta_{\mathrm{rs}} x_\mathrm{s}$, where x'_1 denotes the predicted values from step 1.

The slopes from step 3 are unbiased (we will assume that x_2 through x_s and y are measured without error). However, the standard errors from a two-stage least squares are biased, so special techniques are required. Therefore, it is not a good idea to go through these steps on your own. Rather, use a program like Stata or SPSS that includes a two-stage least squares routine.

An important problem that limits the utility of two-stage least squares is that it is difficult to find instrumental variables that meet the requirements outlined earlier. Most variables that are strongly associated with the explanatory variable are, first, also contaminated by error and, second, associated with the error term. A serious problem arises when the instrumental variables are contaminated by error or do not predict the x variable well: The slopes and standard errors from a two-stage least squares regression model are even worse than the estimates from an OLS regression model.

Although instrumental variables may be of limited utility because they are so difficult to find, let's see an example of a two-stage least squares model—first in Stata (Figure 8.4), then in SPSS (Figure 8.5). The fact that it is hard to find instrumental variables is illustrated by the artificiality of this example. We will use the GSS 2010 data set and make the following (poor) assumptions: (1) health quality (*revhlth*) is measured without error; (2) the true score of occupational prestige predicts health quality, but the observed variable (*prestg80*) is contaminated by measurement error; (3) the two instrumental variables, education (*educ*) and personal income (*income06*), are measured without error; and (4) education and personal income are strong predictors of occupational prestige but are uncorrelated with the error in predicting health quality. In this regression model, we hypothesize that higher occupational prestige is associated

Figure 8.4: Instrumental Variable Regression Output in Stata

```
. ivregress 2sls revhlth (prestg80= educ income06)

Instrumental variabls (2SLS) regression          Number of obs =      1048
                                                 Wald chi2(1)  =     65.50
                                                 Prob > chi2   =    0.0000
                                                 R-squared     =         .
                                                 Root MSE      =    1.0595
```

revhlth	Coef.	Std. Err.	z	P>|t|	[95% Conf. Interval]	
prestg80	.032854	.0040595	8.09	0.000	.0248975	.0408106
_cons	2.155571	.1832164	11.77	0.000	1.796474	2.514669

Instrumented: prestg80
Instruments : educ income06

with better health but that we should use the instrumental variables in a two-stage least squares analysis to reduce the effects of measurement error.

The first step is to estimate the correlations among the explanatory variable and the instrumental variables. In Stata, you can use `pwcorr` to obtain these correlations and their *p* values. In SPSS, you can use *Correlate-Bivariate* to produce these correlations. Figure 8.3 indicates that the variables are moderately correlated, with the largest correlation between occupational prestige and education. Thus, at this point, education and income appear to be reasonable instruments for occupational prestige. Next, we will (in Table 8.1) see the results of the OLS regression model, with life satisfaction as the outcome variable and occupational prestige as the explanatory variable.

The R^2 for this model is .032, with a standard error of the estimate (S_E or Root Root MSE [mean square error]) of 1.026. The explained variance is low but statistically significant ($F = 37.27, p < .001$). The coefficient for occupational prestige is also statistically significant, with, as expected, higher occupational prestige associated with better health. However, we assumed earlier that occupational prestige is measured with error, so let's see what a two-stage least squares regression analysis provides.

In Stata, you can use the instrumental variables command, `ivregress`. This command may also be used to estimate other types of errors-in-measurement models. To request a two-stage least squares for physical health and occupational prestige, include the following on Stata's command line (results presented in Figure 8.4):

```
ivregress 2sls revhlth (prestg80= educ income06)
```

Note that Stata tells us that occupational prestige is "instrumented," with personal income and education as the instruments. Comparing these results to the previous model, note that the coefficient shifts from about 0.013 in the first model to 0.033 in the

Table 8.1: Ordinary Least Squares Regression for Occupational Prestige

Variable	*b*	*SE*
prestg80	0.013	0.002***
intercept	3.032	0.100***
N	1,136	
R^2	.032	
MSE	1.026	

***$p < .001$.

Figure 8.5: Instrumental Variable Regression Output in SPSS

Coefficients						
		Unstandardized Coefficients				
		B	Std. Error	Beta	*t*	Sig.
Equation 1	(Constant)	2.156	.183		11.754	.000
	Prestg80	.033	.004	.445	8.085	.000

two-stage least squares model. Thus, the slope increases substantially after adjusting the explanatory variable for measurement error.

In SPSS, you can estimate the same model by using the drop-down menu *Analyze-Regression-2-Stage-Least-Squares.* The window that opens includes boxes for the dependent, independent, and instrumental variables. The results produced by SPSS are similar to those produced by Stata. The output looks like that presented in Figure 8.5.

How do we know whether this is the proper approach? Although we must rely to large extent on theory (for example, are the instruments well suited to the model?), there is also a statistical test known as the *Durbin-Wu-Hausman* test that allows us to determine whether the instrumental variables approach is preferred. To implement this test, first estimate a linear regression model with the instruments; in the above model, this is:

$$\text{Occupational Prestige}_i = \alpha + \beta_1(\text{Education}_i) + \beta_2(\text{Personal Income}_i) + \varepsilon_i$$

Second, save the unstandardized residuals from this model (`predict res_occ, residual`); in this example, we label them *res_occ*. Third, estimate a linear regression model with the original outcome variable, the original explanatory variables, and the unstandardized residuals. In our example, this model is:

$$\text{Health}_i = \alpha + \beta_1(\text{Occupational Prestige}_i) + \beta_2(res\_occ_i) + \varepsilon_i$$

Figure 8.6: Endogeneity Test Output in Stata

```
estat endogenous

Test of endogeneity
Ho: variables are exogenous

Durbin (score) chi2 (1)      = 35.5134   (p = 0.0000)
Wu-Hausman F(1,1045)         = 36.6538   (p = 0.0000)
```

If the coefficient involving the residuals is statistically significant, then there is evidence that the instrumental variables–two-stage least squares model is preferred to the single-stage linear regression model. In the example of the model designed to predict health, the *p* value of *res_occ* is significant, suggesting that the two-stage least squares model is beneficial compared with the linear regression model.

Stata, but not SPSS, automates this for us, using one of its `estat postcommands`. After estimating the two-stage least squares model, type `estat endogenous`. Stata returns the output shown in Figure 8.6.

Note that the *p* value for these values is similar to the *p* value for *res_occ*. Thus, we reach the same conclusion about the appropriateness of the two-stage least squares model. In general, there is some evidence that occupational prestige is endogenous in this model and that the two-stage regression is preferable to the single regression model (for more information on these tests, see Davidson & Mackinnon, 1993).

Concluding Thoughts

We should regularly ask what our assumptions are before estimating the model. Are such assumptions reasonable? Could there be other potential sources of error in either of the instruments? Suppose we are told that occupations considered prestigious are those that require more education or that pay, on average, higher salaries? Would this change our judgment of education and personal income as instrumental variables? We might decide that these are not good instruments for this model, but finding instrumental variables is typically problematic, and we will always be forced to justify our choices on the basis of reasoning that is divorced from concrete statistical evidence. If we determine that we are not content with the instrumental variables approach, there are other errors-in-variables models that researchers use, but they are beyond the scope of this presentation (if you are a Stata user, you can type `help instrumental` in Stata to see a few of these).

It should be obvious by now that measurement error is an almost universal problem in social science statistics. It may be thought of as one type of specification error in a linear regression model, as it involves extra information (errors in measuring) that we hope to exclude from the model. Whether there will ever be tools to minimize it sufficiently so that we may convince all or most people that our models are valid is, at this time, an unanswerable question. Nonetheless, measurement techniques continue to improve as researchers study the properties of measuring tools and the way that people answer survey questions or report information in general (excellent guides to constructing survey questions are provided by Bradburn, Sudman, & Wansink, 2004; and Krosnick & Fabrigar, 2005). At this point, we should simply do our best to measure social and behavioral phenomena accurately and reliably. Moreover, some useful tools for reducing the damaging effects of measurement error exist and are readily available to even the novice researcher—which is why we recommend taking statistics courses beyond a basic regression course. A review of the literature on factor analysis, multivariate analysis, and structural equation modeling will provide a guide for using these tools.

Key Terms

The following is a list of key terms and concepts in chapter 8. You can find definitions to these words throughout the chapter.

- Factor analysis
- Instrumental variables
- Inverse least squares
- Latent variables
- Measurement error
- Method error
- Multivariate analysis
- Nonstochastic
- Reliability
- Structural equation modeling
- Test–retest reliability
- Trait error
- True score
- Two-stage least squares regression

Analysis Exercise

1. Estimate a model with income (*income06*) as the outcome and occupational prestige (*prestg80*) as the predictor. Treat occupational prestige as a function of instrumental variables for race/ethnicity (*AfricanAm*, *othrace*), gender (*gender*), and education (*educ*). Does the two-stage regression prove to be better than a single regression model?

Solution

1. The model appears to have an endogeneity problem. Our Durbin-Wu-Hausman test for endogeneity was significant, suggesting that the two-stage regression is preferable.

CHAPTER 9 COLLINEARITY AND MULTICOLLINEARITY

Chapter Objectives

In this chapter, we discuss another violation of regression assumptions, collinearity, how it influences regression results, and what can be done about it.

In chapter 9, we will discuss

- collinearity and multicollinearity
- why it is problematic for regression
- how to detect it
- how to reduce the effects of it.

The fifth assumption of linear regression, as we noted in chapter 3, states that there cannot be perfect collinearity among the explanatory variables. The two components of the word *collinearity* are *co-*, or together, and *linear*, or the implication of a straight line or a flat plane. This latter component is tied to the geometric relationship between two variables in space. As we have not discussed the geometric bases of statistical analysis, it is best to refer to collinearity as an issue of covariance. We are interested in the covariability of two or more explanatory variables.

We are not concerned about the covariability of any particular explanatory variable with the outcome variable. In fact, researchers hope to have high covariability between at least one of the explanatory variables and the outcome variable. When we speak about collinearity, we are interested in the covariability among the explanatory variables. Recall that, in a couple of chapters, we have represented covariability by using overlapping circles. The overlap represented statistical control in linear regression in one example and covariability among several variables to demonstrate a latent variable in another. Yet these circles also come in handy for representing collinearity.

Figure 9.1 represents what happens when two explanatory variables are highly correlated. Suppose we wish to determine the unique association between x_1 and y, but x_1 and x_2 have this type of association. Where is the association of x_1 and y controlling for

Figure 9.1: Visualizing Collinearity

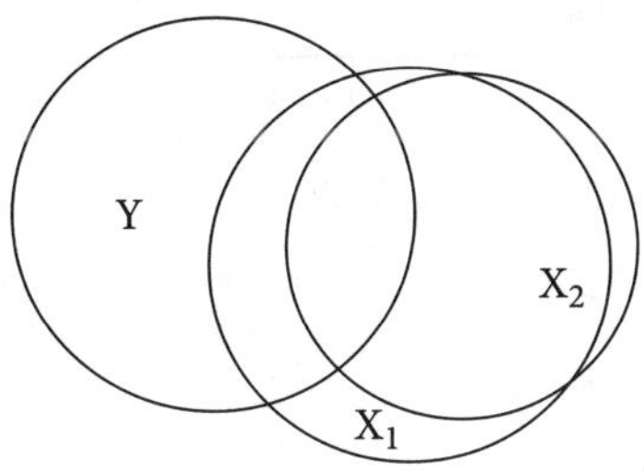

the association with x_2? It is the thin sliver where x_1 overlaps with y but does not include any portion of x_2. However, suppose we wish to estimate this overlap with a linear regression model. Do you see any potential problems? Recall, again, the equation for the standard error of the slope:

$$\text{SE}(\hat{\beta}_i) = \sqrt{\frac{\sum (y_i - \hat{y}_i)^2}{\sum (x_i - \bar{x})^2 (1 - R_i^2)(n - k - 1)}}$$

What will the tolerance $(1 - \text{R}_i^2)$ be if there is such a large overlap between x_1 and x_2? We cannot be certain with a diagram, but rest assured that it will be large. What happens to the standard error of the slope as the tolerance increases? It becomes larger and larger. Suppose, for example, that the correlation between x_1 and x_2 is somewhere on the order of .95. Then, assuming a linear regression model with only two explanatory variables, the tolerance is $(1 - .95^2) = .0975$. Next, take the correlation between another sample's x_1 and x_2 that is much lower, say .25. The tolerance in this situation is $(1 - .25^2) = .4375$. Try plugging fixed values into the standard error equation for the other quantities with these two tolerances, and you will see that standard errors get much bigger when you divide by a small number.

The assumption of ordinary least squares (OLS) regression, though, states that there is no *perfect* collinearity. Perfect collinearity is represented by completely overlapping circles, or a correlation of 1.0. (It could also be −1.0 and have the same consequences.) What happens to the standard error when there is perfect collinearity? The tolerance in this situation is {1 – 1}, which means that the denominator in the standard error equation is zero. Hence, we cannot compute the standard error. It is rare to find a perfectly collinear association, unless you create a variable yourself or accidentally rename a variable and throw it in a regression model with its original twin sibling. Fortunately, statistical software such as Stata and SPSS recognize when two variables are perfectly correlated and throws one out of the model. What is not uncommon, though, is to find high collinearity, such as the figure represents. This is problematic, in part, because

Table 9.1: Ordinary Least Squares Regression of Poor Mental Health

Variable	Coefficient	*SE*	*t*	*p*>*t*
inc	−2.791	1.945	−1.430	0.152
rincom06	2.717	1.951	1.390	0.164
age	−0.052	0.016	−3.250	0.001
Intercept	7.274	0.858	8.470	0.000
N	1,055			
R^2	.020			

statistical software programs do not kick out highly correlated predictors. Thus, it is important to recognize when such problems might be present in your regression model.

The statistical theory literature emphasizes that the main problem incurred in the presence of high or perfect collinearity is biased or unstable standard errors. Slopes are considered unbiased, at least when the model is estimable. Yet practice suggests otherwise. In fact, the slopes of two collinear variables are often incorrect and unstable when two (or more) variables are highly collinear. Let's see an example of this problem.

The GSS 2010 data set has three variables that we suspect are associated with poor mental health (*mntlhlth*): a variable for categorized income (*rincom06*), age (*age*), and year-adjusted income (*inc*). Yet there is something peculiar about these variables: Two of them address income. In most cases, it is probably unlikely that a researcher would do this, but for the purposes of showing the effect of collinearity, we will do it here. First, let's consider the correlation between *income06* and *inc*. Both Stata and SPSS tell us that the correlation between the two is .9999 ($p < .001$), which would be considered a very strong correlation. Hence, we have two variables with much of their variability overlapping.

We will estimate a linear regression model using the two income variables and age as predictors of mental health, which we enter as the outcome variable. Depending on whether you used Stata or SPSS, you should get results that look something like the results shown in Table 9.1.

Look at the unstandardized coefficients for the two variables: One is negative and one is positive, although they largely measure the same thing. In some cases, two collinear variables will have two different signs (one will be positive, while the other will be negative). Although statistical theory suggests that these slopes are unbiased, they are untenable. How could two related variables go in two different directions?

Another warning sign is when coefficients and standard errors shift quite a bit from one model to the next. For example, we could imagine a regression model for state-level homicide rates where two predictors are collinear. Imagine if we excluded

Table 9.2: Ordinary Least Squares Regression of Poor Mental Health (with Collinearity Removed)

Variable	Coefficient	*SE*	*t*	*p*>*t*
income06	−0.081	0.035	−2.340	0.019
age	−0.054	0.016	−3.290	0.001
Intercept	7.375	0.856	8.620	0.000
N	1,055			
R^2	.018			

Utah from the next model we ran, and the coefficients and standard errors for the two collinear variables have shifted quite a bit from one model to the next. Although this may be because the state of Utah has unusual values on one or more of the variables in the model (see chapter 12), it might also be the result of collinearity between the two variables. In fact, it is not uncommon for collinearity to create highly unstable coefficients that shift substantially with minor changes in the model.

Of course, it is unlikely that we would have estimated a model with two variables that are so similar. Even though we used these models to illustrate a point, it is important to always consider whether two (or, as we shall see, more) explanatory variables are highly correlated *before* estimating the model. Now let's see what happens when we include only one income variable in the OLS regression model that predicts mental health.

The results of this model (see Table 9.2) show that there is a negative association between poor mental health days and income after adjusting for the association with age. It is important to note how different the coefficient is from those in the earlier models. More important, notice the standard error for *income06* in this model: It is much smaller than the standard errors from the earlier models. This demonstrates how collinearity inflates standard errors substantially.

Multicollinearity

Recall that the tolerance from the auxiliary regression equation is based on the proportion of variance among one explanatory variable that is explained by the other explanatory variables. In other words, it is the R^2 from a linear regression model. We hope it is clear by this point that it is important to consider overlapping variability between not only two variables but among all the explanatory variables. Suppose, for example, that we have four explanatory variables, x_1, x_2, x_3, and x_4. The largest bivariate correlation between any two variables is .4. If we were to use only the two variables with the largest

Figure 9.2: Visualizing Multicollinearity

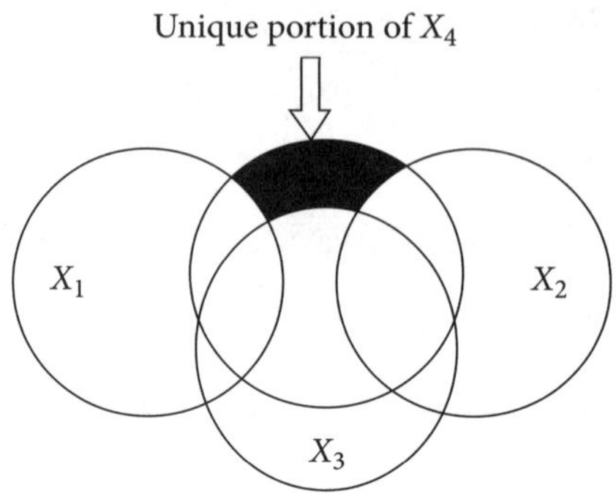

correlation in a linear regression model, the tolerance in the standard error formula would not be any smaller than $(1 - .4^2) = .84$. It is unlikely that a tolerance of this magnitude would bias the standard errors to an extreme degree or make the slopes unstable. However, it is possible for one of these explanatory variables to be a *linear combination* of the others. If, for instance, x_4 is predicted perfectly or nearly perfectly by x_1, x_2, and x_3, then the tolerance will approach zero and the same problems result: biased standard errors and unstable slopes. In terms of our overlapping circles (see Figure 9.2), multicollinearity appears as two or more variables encompassing another explanatory variable. It should be clear, though, that looking at bivariate correlations will not always be sufficient for determining whether multicollinearity is present. Any of the bivariate correlations may be below thresholds that warn of collinearity, but they won't show if some combination of explanatory variables predicts another explanatory variable. We must use other diagnostic tools.

How to Detect Multicollinearity

As we have already mentioned that bivariate correlation matrices are of limited use for detecting multicollinearity (although they are useful for determining collinearity), we should find some other diagnostic tools. First, though, it is worth mentioning a rule of thumb that has been used to gauge collinearity: Estimate a correlation matrix and then look for bivariate correlations among the explanatory variables that exceed .7. Any correlation that matches or exceeds .7 should lead you to suspect that collinearity may affect your results. (This depends on the sample size, however.)

What about multicollinearity? Is there a test that will allow us to determine its presence? How would one go about devising such a test? You were given a hint when we discussed the tolerance formula in the standard error equation. In chapter 3, we mentioned that the tolerance was based on the auxiliary regression model that includes x_i as the outcome variable and all the other explanatory variables as predictors (for example, $x_{1i} = \hat{\alpha} + \hat{\beta}_2 x_{2i} + \cdots \hat{\beta}_k x_{ki}$). Can we use this information to assess the possibility of multicollinearity? The easy answer is yes—we may simply regress each explanatory variable on all the others and compute an R^2 for each model. We need a rule of thumb, though: How high must an R^2 be before we suspect multicollinearity? Perhaps we should use the same rule of thumb as mentioned earlier: .7. However, because we square the R, perhaps it should be .49. Although there are rules of thumb that specify the threshold for the auxiliary R^2 and the tolerance, a more frequently used statistic is known as the *variance inflation factor* (VIF). For various reasons we won't discuss here, the VIF has become the test du jour for assessing multicollinearity. It is defined as the inverse of the tolerance:

$$\text{VIF} = \frac{1}{\{1 - R_i^2\}}$$

The rule of thumb is that a VIF greater than or equal to 10 (some say it is nine) is indicative of multicollinearity. Others argue that if the square root of the VIF is greater than or equal to three, we should suspect multicollinearity. A relevant issue that is not often discussed, though, is sample size: Linear regression models are affected less by collinearity or multicollinearity as the sample size increases. Hence, all else being equal, a model that uses a sample size of 10,000 is less sensitive to multicollinearity—at least using conventional rules of thumb—than a model that uses a sample size of 100. Unfortunately, the decision rules that have emerged for these tests do not normally consider the sample size.

Another standard test for multicollinearity involves *condition indices* and the *variance proportions*. Before demonstrating how to use this test, it is useful to take a quick journey into the intersection of linear algebra and principal component analysis for some terminology. An *eigenvalue* is the variance of the *principal component* from a correlation (or covariance) matrix. The principal components of a set of variables are a reduced set of components that are linear combinations of the variables. They are one type of latent variable (see chapter 8), yet the principal components from a set of variables are uncorrelated with one another. Hence, if we have a set of, say, 10 explanatory variables, we may reduce them to a set of, perhaps, two or three principal components that are uncorrelated with one another. The eigenvalues of these principal components measure their variances. Briefly, then, eigenvalues are a measure of joint variability among a set of explanatory variables. As the eigenvalue gets smaller, there is more shared variability among the explanatory variables, with a value of zero indicating perfect collinearity. An interesting property of eigenvalues in the following test is that they sum to equal the number of predictors in the regression equation (number of explanatory variables + the intercept, or $k + 1$).

How is this useful for judging multicollinearity? We may use eigenvalues to compute condition indices, which are another measure of joint variability among explanatory variables. Condition indices are used in conjunction with *variance proportions* (Stata calls these *variance-decomposition proportions*) to assess multicollinearity in linear regression models (they may also be used in other regression models). Variance proportions assess the amount of the variability in each explanatory variable (as well as the intercept) that is accounted for by the dimension (or principal component). Fortunately, statistical programs like SPSS and Stata will compute these values for us. Let's now move onto an example of VIFs, eigenvalues, condition indices, and variance proportions. The manner by which the collinearity tests in SPSS and Stata are conducted are quite different, so we present them separately below.

Stata

In Stata, we begin by estimating the linear regression model we used earlier that included both income variables, then use the `vif` command to calculate the VIFs. Next, use the `coldiag2` command and the `collin` command followed by the explanatory variables (`collin inc income06 age`) to see additional collinearity diagnostics. To use the `coldiag2` and `collin` commands, we may need to download files from a Stata server. Assuming we are connected to the Internet, we may type `findit coldiag2` and `findit collin` to locate the files.

Entering `vif`, `coldiag2`, and `collin` on the command line after regress provides the information in Figures 9.3 and 9.4. The collinearity statistics reported in Figure 9.3 show the VIFs and the tolerances (1/VIF). You can see that two of these obliterate the rule of thumb that any VIF of 10 or more indicates collinearity or multicollinearity problems, but we may also use the condition indices and variance proportions in the third table to assess collinearity. Here is how it's done. First, look for any condition indices that are greater than or equal to 30. Then look across the row and find any variance proportions that exceed .50. A condition index that is greater than or equal

Figure 9.3: Variance Inflation Factor (VIF) Test Output in Stata

vif

Variable	VIF	1/VIF
inc	3284.53	0.000304
rincom06	3284.06	0.000305
age	1.03	0.972315
mean VIF	2189.87	

Figure 9.4: Collinearity Diagnostics in Stata

. collin inc rincom06 age
(obs = 1345)

Collinearity Diagnostics

Variable	VIF	SQRT VIF	Tolerance	R-Squared
inc	4765.14	69.03	0.0002	0.9998
rincom06	4765.77	69.03	0.0002	0.9998
age	1.04	1.02	0.9594	0.0406

Mean VIF 3176.98

	Eigenval	Cond Index
1	3.7763	1.0000
2	0.1734	4.6666
3	0.0503	8.6678
4	0.0000	461.6565

Condition Number 461.6565
Eigenvalues & Cond Index computed from scaled raw sscp (w/intercept)
Det (correlation matrix) 0.0002

to 30 coupled with a variance proportion that is greater than or equal to .50 indicates a collinearity or multicollinearity problem that should be checked further. An advantage of condition index and variance proportions over VIFs is that they identify which variables are highly collinear.

In the collinearity diagnostics shown in Figure 9.4, there is one condition index, associated with line 4, which exceeds 30; it is 731.19 or 1,193.40 (there is slightly different scaling for `coldiag2` and `collin`). Combined with our VIF, we have strong evidence that these two variables are highly collinear.

SPSS

In SPSS, we can use the drop-down menus *Analyze-Regression-Linear* and then click on the box labeled Statistics. Next, place a check in the box labeled Collinearity diagnostics. Then, estimate the linear regression model with *inc*, *income06*, and *age* as predictors of mental health. The results appear in Figure 9.5.

Notice that the collinearity in our regression is so bad that it drops the variable *inc* from the model; however, we see that the VIF for the variable is 15,747.19, which is substantially larger than the VIF cutoff of 10, again suggesting that there is substantial collinearity in the model.

Figure 9.5: Collinearity Diagnostics and Variance Inflation Factor (VIF) Test in SPSS

Coefficients[a]

Model	Unstandardized Coefficients		Standardized Coefficients	t	Sig.	Collinearity Statistics	
	B	Std. Error	Beta			Tolerance	VIF
1 (Constant)	8.152	.930		8.770	.000		
TOTAL family income	–.135	.041	–101	–3.259	.001	.955	1.047
AGE of respondent	–0.45	.016	–.085	–2.744	.006	.955	1.047

[a]Dependent Variable: days of poor mental health past

Excluded Variables[b]

Model	Beta In	t	Sig.	Partial Correlation	Collinearity Statistics		
					Tolerance	VIF	Minimum Tolerance
1 inc	4.715[a]	1.238	.216	.038	6.350E-5	15747.190	6.350E-5

[a]Predictors in the Model: (Constant), age respondent, total family income
[b]Dependent Variable: days of poor mental health past

What Can We Do about Collinearity and Multicollinearity?

Now that we have learned how to detect these problems, we need to know about some possible solutions. We mentioned the first solution earlier. Collinearity and multicollinearity have less influence on linear regression models as the sample size gets larger. Therefore, if a model is plagued by collinearity problems, then, if possible, one should add more observations. Unfortunately, this is rarely feasible outside of a laboratory. Survey data are typically of a fixed sample size because of cost constraints, so adding more observations is not practical.

Another solution was discussed in chapter 8. Suppose that some subset of explanatory variables is involved in a multicollinearity problem. Look over these variables: Are they measuring a similar phenomenon? Would it be practical to combine them in some way? Often, highly collinear variables are indirectly measuring some latent construct, so using them to create a latent variable is a common solution to multicollinearity. Rather than including the collinear variables in the regression model, we may place the latent variable in the model. Then we may assess its association with the outcome variable. For example, suppose we have variables that are designed to measure sadness, happiness, melancholy, depression, and grief. When we try to place all of these variables in a model designed to predict number of sick days in the last month, we find that there are substantial multicollinearity problems, and we find that they are all highly correlated, with each bivariate correlation about .8. In this situation, it might be a good idea to combine the variables by using a latent variable technique, such as factor analysis, and consider what this latent variable is actually measuring. It might be measuring, for instance, a general trait of sadness or general feelings of depression. This likely helps to account for the number of sick days a person takes in an average month.

The third solution is to use a technique known as *ridge regression*. Although it is rarely recommended as a general solution, ridge regression may offer some help when nothing else seems to work or in particular situations that we won't cover here. The general idea underlying ridge regression is to create (artificially) more variability among the explanatory variables. This is usually done by considering, rather than the raw data, the variance–covariance matrix of the variables. The diagonals of this matrix are the variances, which are too small in the case of collinearity or multicollinearity to yield unbiased estimates of standard errors. Hence, in ridge regression, the analyst chooses a constant to add to the variances, thus increasing variability and decreasing the effects of multicollinearity. The choice of this constant may be arbitrary or based on a specific formula. This may sound very promising, but you should realize that ridge regression results in biased slopes. However, the bias may be less than the OLS estimates that are made unstable by multicollinearity. A key problem with ridge regression, though, is that it seems to change slopes that are not significant in an OLS model more than slopes that are significant. Therefore, the analyst must have solid conceptual hypotheses to determine which slopes are important and which are not (see Draper & Smith, 1998, for a review of ridge regression).

The fourth solution, which is not normally recommended, is to use a variable selection procedure, such as stepwise or backward elimination, to choose the "best" set of predictive variables (see chapter 7). These methods typically will drop one or more of the variables involved in the collinearity or multicollinearity problem. However, they also take the important conceptual work out of the hands of the analyst and present other problems that suggest they should be avoided.

The last solution is recommended the least often but is probably used most often by analysts who use regression techniques. This is to simply omit one or more of the variables involved in the collinearity or multicollinearity problem. This is recommended only as a last resort or if one of the collinear variables measures virtually the same phenomenon as other collinear variables. It is rare, however, for two variables to measure the same thing. It is much more common to find that collinearity is caused by recoding problems in the data cleaning stage. This may occur when one variable is created from another.

Generally, it should be clear that collinearity and multicollinearity present some potentially serious problems for the linear regression model. Biased standard errors are a nuisance, but, when coupled with unstable slope coefficients, the results of the model are not trustworthy. What should you do about this problem? It is best to consider carefully the source of collinearity or multicollinearity and then use the best solution available. Unfortunately, there are no easy answers.

Key Terms

The following is a list of key terms and concepts in chapter 9. You can find definitions to these words throughout the chapter.

- Collinearity
- Condition indices
- Eigenvalue
- Linear combination
- Multicollinearity
- Principal component
- Ridge regression
- Variance inflation factor (VIF)
- Variance proportions

Analysis Exercises

1. Estimate a model with mental health (*mntlhlth*) as the outcome and age (*age*), income (*income06*), year of birth (*cohort*), gender (*gender*), and race (*AfricanAm, othrace*) as predictors. Then run a VIF test to address the possibility of multicollinearity. What does the VIF test show?
2. Estimate a model with income (*income06*) as an outcome that is predicted by gender (*gender*) and race (include *AfricanAm, othrace,* and *Caucasian*). What happens to the model?

Solutions

1. Age and cohort are highly collinear. The VIF for age is 1,452.21, and the VIF for cohort is 1,450.01. The correlation between the two variables is –.9981, which provides further evidence that the two are measuring the same thing.
2. By including all three racial/ethnic groups, Stata and SPSS will drop one of the groups (usually randomly); this is because there is perfect multicollinearity among the three dummy variables measuring race/ethnicity.

CHAPTER 10 NONLINEAR ASSOCIATIONS AND INTERACTION TERMS

Chapter Objectives

In this chapter, we will discuss special functional forms of variables, nonlinear variables and interacted variables, which are commonly used in linear regression models.

In chapter 10, we will discuss

- when to use nonlinear terms, such as quadratic functions and logged variables
- how to tell whether a nonlinear term works or is useful
- the value of interaction terms in regression models
- how to interpret nonlinear and interaction effects in regression models.

In chapter 3, we discussed an important assumption of the linear regression model: The mean value of *y* for each specific combination of the *x*s is a linear function of the *x*s. We also saw briefly at the beginning of chapter 7 that one type of specification error involves this assumption. When the association between *y* and one of more of the *x*s is not represented well by a straight line or flat surface, then we have one form of specification error. A linear regression analysis is going to attempt, using least squares techniques, to fit a straight line to a set of data. However, as shown in Figure 7.1 (which is reproduced as Figure 10.1), it is clear that when the association between *x* and *y* is not linear—rather, when it is curved or *curvilinear*—the OLS estimates will not yield the best results.

Curvilinear and other associations among variables that are not represented well by a straight line are known generally as *nonlinear* associations. Although making comparisons across the social and behavioral sciences would be very difficult, it is likely that nonlinear associations are more normative than we think. Yet linear associations are much simpler to conceptualize and model. Many statistical procedures, like correlations and OLS regression analysis, are built on the assumption of linearity. Although there is a growing literature on nonlinear regression techniques, the way we think about associations among variables still tends to be linear. This may have something to do with general limitations of the human mind. Thankfully, statistical computer programs, like

Figure 10.1: Example of a Nonlinear Relationship

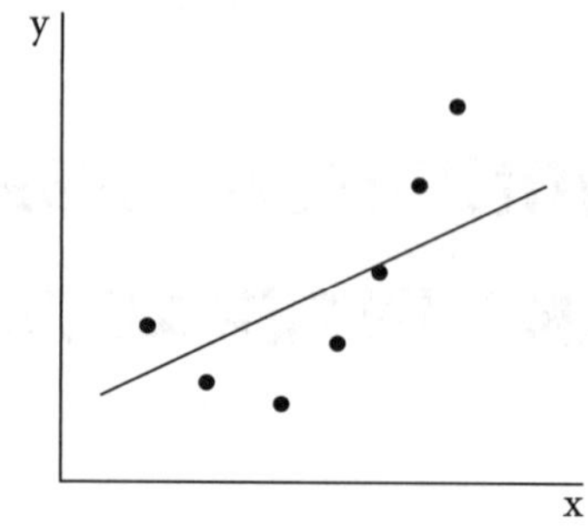

Stata and SPSS, are quite capable of modeling nonlinear relationships. Nevertheless, it is likely that we simply need more experience with nonlinear thinking and modeling.

There are actually two approaches to nonlinear analysis that are used in regression modeling. We will begin with the more complex approach first, followed by the simpler nonlinear relationship. The first type of nonlinear regression analysis proposes that the estimated coefficients are nonlinear; in other words, they do not imply a straight line pattern. An example of a model that is nonlinear in the estimated parameters is:

$$y_i = \alpha + \beta_1 x_{1i} + \beta^2{}_1 x_{2i} + \log_e(\beta_2 x_{3i}) + \beta_3 x_{4i} + \varepsilon_i$$

In this equation, the betas are not simple, linear terms; rather, they include *higher-order terms*, with nonlinearities such as quadratics (for example, squaring a coefficient) and logarithms ($\log_e$ implies the natural, or Naperian, logarithm). There are nonlinear regression routines that are designed specifically for situations when the estimated parameters are hypothesized to be nonlinear. However, such hypotheses can be difficult to formulate, so we will not cover this type of nonlinear regression approach here. Another promising avenue in regression analysis, but one that we also will not discuss here, is known as *nonparametric regression*. An example of this model is known as the generalized additive model, which allows the analyst to fit various types of associations—linear and nonlinear—among the outcome and explanatory variables (see Fox, 1997, for a good overview of such models).

In this chapter, we will address a simpler approach that adapts the linear regression model so that it may incorporate nonlinear associations among the explanatory and outcome variables. To begin to understand how we may do this, consider that when we estimate a linear regression model and interpret the slopes, we assume that the magnitude of change in y that is associated with each unit change in x is constant. In other words, it is the same, regardless of where in the y distribution the change is presumed to take place. For example, suppose we find, as in chapter 2, that the slope of education on poor mental health days is −0.275. By now, you should be able to interpret this

Figure 10.2: Linear Regression Example

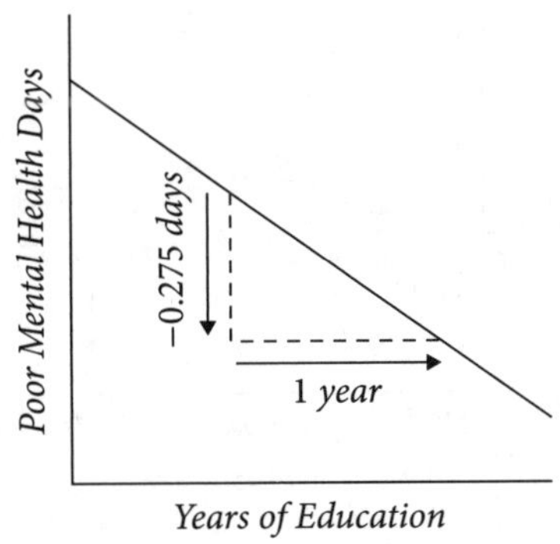

coefficient: Each one-year increase in education is associated with a 0.275-day decrease in the number of poor mental health days. Keeping this in mind, consider Figure 10.2.

A linear relationship assumes that the association between education and poor mental health is the same, regardless of where we are on the educational continuum. Of course, as mentioned in chapter 2, we should interpret these associations only within the bounds of our data. Suppose, though, that when individuals are less educated, say, with 11 years or fewer of education, a year increase in education is *positively* associated with poor mental health days, whereas highly educated people show a decrease in poor mental health days. Perhaps highly educated individuals have better access to mental health services, are less worried about meeting their temporal needs, or have creature comforts that improve their outlook on life. Whatever the hypothesized mechanism, these sorts of possibilities are legion and recommend against simple hypotheses that suggest only linear associations.

Let's look at an example of a nonlinear association. In the GSS 2010 data set, there are variables that measure personal income (*rincom06*) and the respondent's age in years (*age*). Before analyzing these data, think for a moment about people's incomes in the United States. Among adults, how should the association between age and income behave? First, it seems clear that younger adults who are still in school or just entering the workforce have less income than middle-aged or older adults who have been in the workforce longer (we will also ignore for now the influences of education, unemployment, job training, and other important variables). What happens as people get older and reach their 60s? Many begin to retire, and their personal income tends to decrease. Thus, we have already departed from hypothesizing a linear association between age and personal income. Rather, there is likely a curvilinear association, with income rising with age until it reaches a maximum point, after which it begins to decrease as people retire. However, there are also important demographic issues to consider when dealing with age. For example, wealthier people, on average, tend to live longer than poorer people partly because they have access to better health care. As we begin to look at the age–income

Figure 10.3: Associations between Income and Age

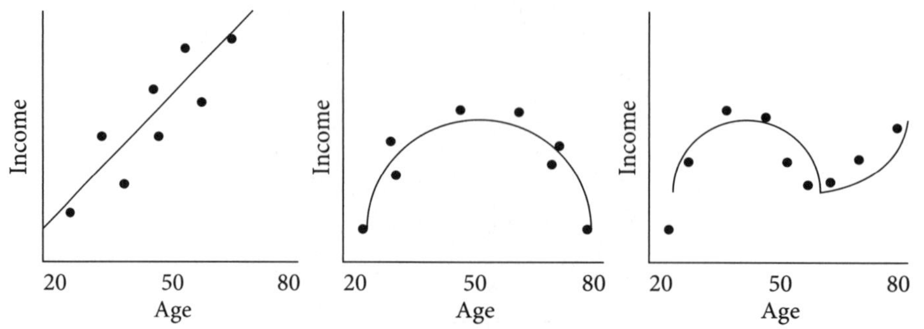

association at older ages, say 75 or 80, we may begin to see the effects of certain people dying earlier than others. If, on average, those with less income tend to die earlier, then income should appear higher among older adults. There are many complex demographic phenomena at play here that we won't address here, but we should consider that the age–income association is not linear and may have more than one "bend."

The graphs shown in Figure 10.3 represent the three associations between age and income that we have discussed thus far. The left graph shows a linear association, with income increasing at a steady pace with increasing age. The middle graph shows what is known as a *quadratic* association, with one "bend" in the association. The right graph shows a *cubic* association, with two bends—and three distinct pieces—to represent the last part of our discussion about age and income. Once we begin dealing with nonlinear associations, they can quickly blossom into some very complex shapes with numerous bends and twists. Therefore, to manage the modeling, it is important to always consider conceptually the associations in your set of variables. As we saw in earlier chapters, there are routinized ways to estimate regression models (for example, stepwise selection, reliance on nested *F* tests), but these should be avoided if they are not guided by theoretical considerations.

When thinking about associations among any set of variables, it is wise to consider whether there are nonlinear associations. You may be surprised at how often they occur. For instance, the association between age and various outcomes is often curvilinear. The impact of explanatory variables such as education on outcomes often have what are known as *floor* or *ceiling effects*: There is a negative or positive association up to a certain point; thereafter, there is a flat association.

Figure 10.4 shows an example of a hypothetical ceiling effect concerning the association between peer delinquency and one's own delinquency. As the number of friends involved in delinquency increases, one's own delinquency also tends to increase up to a certain point. At that point, high-frequency delinquents, though, are often psychologically different from others (for example, more impulsive, less empathetic, more callous), so having more delinquent friends may not add much to one's own delinquency.

Figure 10.4: Visualizing a Ceiling Effect

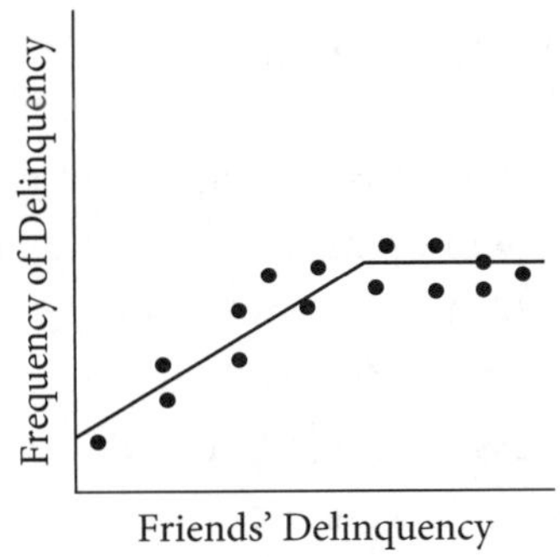

If you suspect that there are nonlinear associations among your variables, there are some simple tools for checking them. Perhaps the most useful is the ordinary scatter plot. Returning to our income and age example, we will use the GSS 2010 data set to construct a two-variable scatter plot with personal income on the *y*-axis and age on the *x*-axis.

Stata

Let's create the scatterplot in Stata by using the command: `scatter rincom06 age, jitter(5)`. Although we have looked at scatter plots before, the jitter option is new; it randomly "jitters" the data points on the graph so that any that may overlap will be easier to see. The result is very busy as there are so many data points. It is impossible to visualize whether a straight line or a curved line fits this association best. Fortunately, Stata provides additional options to fit a variety of lines to its plots: the mean of *y*, a linear fit line, a quadratic fit line, and a lowess fit line. The mean of *y* fits a horizontal line; `twoway scatter rincom06 age, jitter(5) yline(10.35)`. A linear line places the regression slope in the plot (`twoway lfit rincom06 age`), a quadratic line fits a line with up to one bend (`twoway qfit rincom06 age`), and a lowess fits locally weighted regression lines that allow a series of lines to fit the data points (`twoway lowess rincom06 age`). The last option parcels the data into smaller (local) sections and fits regression lines to each section. This is useful when we are not sure what type of nonlinearity might occur.

SPSS

Running an analysis similar to that in Stata is also possible in SPSS by using its graphing options. To do so, use the drop-down menu *Graph-Scatterplot-Simple* to construct a two-variable scatter plot with *income* on the *y*-axis and *age* on the *x*-axis. Like the original graph we produced in Stata, the SPSS graph is very difficult to read. Thankfully, we can use some of SPSS's options to make the scatter plot easier to interpret. Double clicking on the scatter plot and highlighting the data points, which will produce a set of icons on the top of the window, can accomplish this. Place your mouse over the icon that looks like a scatter plot with a straight diagonal line. It should read, "Add fit line." Click on this icon once, which will open the Properties window. This window shows you the five lines that are available to you: the mean of *y* (a horizontal line), a linear fit line (OLS regression line), a quadratic fit line (a line with one bend), a cubic fit line (a line with two bends), and a lowess fit line (a line with varying bends).

Whether you use SPSS or Stata, you should try alternating between linear, quadratic, and lowess to determine the estimated association between age and personal income. It is interesting to look at the types of associations that emerge. In particular, although it is difficult to see in any of the plots, there is a quadratic association up until about age 60, and then the curve declines. Hence, there is evidence that the association between age and personal income follows a squared pattern. If it followed a linear pattern, the linear, quadratic, and lowess would yield the same straight line.

Partial Residual Plots

Of course, we may not always know whether the model includes these types of associations. Therefore, we will now see a way to test for nonlinearities in a linear regression model and explore what to do when faced with a nonlinear association. We will stick with the same simple example but specify *personal income* as the outcome variable and *age* as the explanatory variable. If you are using Stata, you can estimate this model, then use the following postregression command to predict the studentized residuals: `predict rstudent, rstudent`. In SPSS, this can be achieved by clicking the Save button and checking the Residuals-Studentized box before running the OLS regression. Overall, the regression results indicate that there is a positive association between age and personal income, but this does not examine the nonlinear association between these two variables, which we now know exists. By constructing a scatter plot with the studentized residuals (which scale the residuals by using a *t* distribution) on the *y*-axis and age on the *x*-axis, however, we can learn something (in Stata, this can be done with the commands `twoway scatter rstudent age`; in SPSS, you can create a scatter plot with the residuals on the *y*-axis and age on the *x*-axis). This type of scatter plot is known as a *partial residual plot*. Many analysts recommend that we view a partial residual plot for each explanatory variable used in a regression model. Keep in mind that, in our example, the residuals gauge all the influences on personal income except for its *linear* association with age. Based on an assumption of the regression model, there should be no association between age and these residuals—but is there? We can figure this out by plotting a fit line in the partial residual plot you have just created. Try out whichever type of fit line you would like, but be advised that many experts recommend using the lowess line because it will show many possible associations. If you choose any of the nonlinear fit lines, you will notice the curvilinear association between age and the residuals. This is indicative of a nonlinear association that we haven't considered in our regression model, but, as we suspect, it is the *nonlinear* aspects of age and personal income that are showing up.

Unlike SPSS, Stata can automate this procedure by using the postregression command `cprplot` (component-plus-residual plot). By typing `cprplot age` after the original model, you will find a scatter plot that looks similar to the one using two-way `scatter rstudent age`. However, notice that `cprplot` adds a linear fit line. Moreover, adding the subcommand `lowess` (`cprplot age, lowess`) provides a lowess fit line that demonstrates the nonlinear association between age and personal income.

Now that we know a bit more about the association between age and personal income, what should we do about it? There are a couple of approaches that are particularly helpful. This involves a linear regression model with *higher-order terms for age*. Using the graph we produced earlier, we should create a squared term for age. This means that we take age and square it (age × age; or age^2 [age raised to the power of 2]). If you are using Stata, you may create these variables using the generate command (`generate age2 = age*age` or `generate age2 = age^2`). The names for these new variables are arbitrary but indicate what they represent. Next, estimate a regression model that includes *rincom06* as the outcome variable and *age* and *age2* as the explanatory variables. In SPSS under *Analyze-Regression,* you can use an option labeled Curve Estimation. Choose this and then place *personal income* in the box labeled Dependent(s) and *age* in the box labeled Independent. You will notice at the bottom of the window a set of boxes under the heading Model. By checking these boxes, you can fit a host of nonlinear associations. For now, let's stick with linear and quadratic. After clicking *OK,* you'll find output consisting of three regression models. In the interest of space, we will look only at the model we ran in Stata and the final model provided by SPSS.

Table 10.1 shows two explanatory variables, the latter corresponding to the quadratic version of age. Both the linear and quadratic terms are significant. Moreover, the signs of the slopes tell us something important (although we already knew this) about the nonlinear association between age and personal income. The signs are positive and negative, corresponding to an initial positive association and a subsequent negative association. It is also quite simple to determine at which ages the association turns negative and then positive. We will see one way to do this later.

A question to ask is whether this model is an improvement over models with just age or with age, age^2, and age^3 as explanatory variables. One simple way to determine

Table 10.1: Personal Income Regressed on Linear and Nonlinear Terms for Age

Variable	*b*	*SE*
age	0.969	0.065***
age2	−0.010	0.001***
Intercept	−7.735	1.420
N	1,201	
R^2	.169	

***$p < .001$.

this is to compare the adjusted R^2 values from the three models and combine them with nested F tests to determine statistical significance (see chapter 5), but we first need to learn a bit more about how to estimate the model.

It is always a good idea when using explanatory variables with higher-order terms (for example, age *and* age^2) to check for multicollinearity among the variables. Let's do this and see what the results show. The VIFs for the explanatory variables are very high: 36.62. There is—and this should not be surprising—substantial multicollinearity in this model. Another symptom of multicollinearity is the large standardized regression coefficients; it is rare to find one larger than 1.0, yet, using the beta subcommand, we can see that all three are much larger than one (in absolute value). Recall, though, that multicollinearity inflates standard errors, making it harder to find significant results. Yet both age variables have significant slopes, supporting the hypothesis of a quadratic association between age and personal income.

Nonetheless, most of us do not care for such extreme multicollinearity in our linear regression models, but is there anything we can do about it? There is a simple statistical trick that allows us to diminish the multicollinearity problem. It is based on the following statistical phenomenon:

$$\mathrm{Cov}(x_i - \overline{x}, \overline{x}) = 0$$

You may substitute correlation for covariance, and the results will be the same. The proof of this statement is provided in many books on probability or statistical theory, but outside of what we will cover here. What it means for us is that by centering the explanatory variable before computing the higher-order terms, and then using these new terms in the linear regression model, multicollinearity is reduced substantially. You may recall that perhaps the most common form of centering a variable is to compute its z scores.

Table 10.2: Personal Income Regressed on Linear and Nonlinear Terms for Standardized Age

Variable	*b*	*SE*
Zage	0.726	0.209***
Zage2	−3.022	0.219***
Intercept	16.488	0.224
N	1,201	
R^2	.169	

***$p < .001$.

Stata

If you are using Stata, the simplest way to compute *z* scores is with the `egen` command: `egen zage = std(age)` (there is also a downloadable command, `zscore`, that may be used). Using this command, we find that the variable *zage* (though we could call it anything we'd like) has been added to the data set: It consists of the *z* scores for age.

Next, let's use *zage* to create a new quadratic age variable. In Stata, the following line produces this variable: `generate zagesq= zage^2`. After creating these variables, estimate a linear regression model with *personal income* as the outcome variable and *zage* and *zagesq* as the explanatory variables. Make sure you ask for collinearity diagnostics after the model. The results shown in Table 10.2 should be listed in Stata's output window.

SPSS

In SPSS, the simplest way to compute *z* scores is with the *Analyze-Descriptive Statistics-Descriptives* drop-down process. Place the variable in the Variable(s) box, and check the box in the lower left-hand corner that reads "Save standardized values as variables." Going through these steps with age, we find that the variable *Zage* has been added to the data set. This variable consists of the *z* scores for age. Now, let's use *Zage* to create a new quadratic variable. In the SPSS syntax file, you can do the following to produce such a variable:

```
COMPUTE zagesq= Zage **2.
EXECUTE.
```

Keep in mind that even though the age variables are now measured in different units, the results are quite similar to the earlier results: a positive association that turns negative (as shown in Table 10.2). More important, though, the VIFs are well below the

Figure 10.5: Nonlinear Relationship Graph Produced by Stata

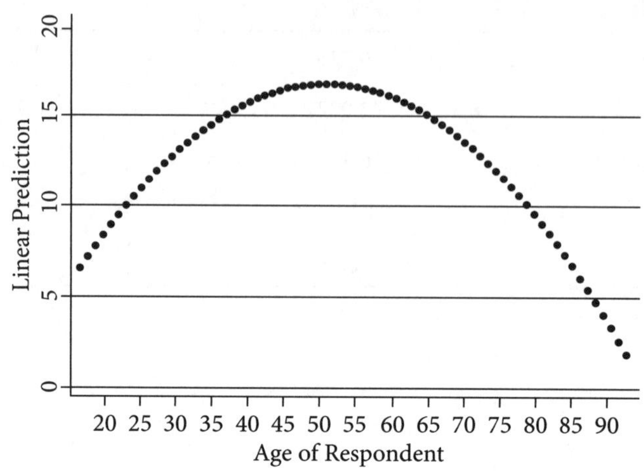

thresholds indicative of multicollinearity. It is a good idea to remember this statistical trick for dealing with collinearity and multicollinearity problems; it will come in handy later in the chapter and perhaps throughout your career in regression modeling.

A problem with using these centered explanatory variables emerges if we wish to determine at what point in the age or personal income distribution the slope turns negative or positive. One valuable approach uses a scatter plot. This scatter plot uses the unstandardized predicted values (in Stata, these are predicted by using the `predict` command with the `xb` option: `predict pred, xb`; in SPSS, these can be saved by checking the appropriate box in the Linear Regression Save window) as the y-axis and *age* as the x-axis. Figure 10.5 shows that the initial downturn in the association occurs close to age 55 (use either Stata's or SPSS's graph editor to change the scale markers of the x-axis from 20 to 5 to help you visualize this), but consider the following question: Why is the downturn at such a relatively young age? Shouldn't it occur much closer to the typical retirement age? (Hint: It might have something to do with the nature of cross-sectional data and birth cohorts.)

Nonlinear associations abound in the social and behavioral sciences (as well as in other scientific disciplines, such as chemistry, biology, and physics). We have seen only a couple of simple examples, but when we consider variables such as age, education, socioeconomic status, and many other interesting variables, we often find nonlinear shifts in their associations with numerous outcome variables. It is best to first think carefully about potential nonlinear associations among your variables (does the literature suggest any?) and then use visual techniques, such as scatter plots, to examine them. Then, once you have a sense of the likely shape of the associations, consider ways to test for them in a linear regression model. Squaring or cubing explanatory or—as we shall see—outcome variables scarcely uncovers the myriad possibilities.

Is the Error Term Normally Distributed?

Another assumption that falls under nonlinearities (although indirectly) involves the outcome variable. As discussed in chapter 3, we assume that the error term is normally distributed. Recall that the error term (or residuals) measures everything about the variability of the outcome variable once we have taken into account the part of it that may be predicted by the explanatory variables. We hope that all that is left is random noise, but this is not always the case. Nonetheless, we still assume that the part that is left over is normally distributed; in other words, it follows a normal (Gaussian) distribution. It should be clear that this assumption is examined after estimating the regression model, even though it is common to see analysts assess whether the outcome variable itself is normally distributed. Because the latter step is so frequently taken, let's see how we might examine the "normality" of the outcome variable.

The most convenient way to examine the distribution of the outcome variable is with a *normal quantile–quantile* or *normal q-q plot.* When using this plotting approach to test the assumption of normally distributed errors, some statisticians call it a *normal probability plot* or a *Gaussian rankit-plot.* It is rare in the social and behavioral sciences that an outcome variable is continuous; it is rarer still to find one that is normally distributed. Because this book deals with continuous outcome variables (but see chapter 13), we will ignore the issue of noncontinuous outcome variables (for more information, see Hoffmann, 2004). A normal *q-q* plot provides a straightforward way to assess whether a variable is normally distributed. Here is how it works. First, the statistical program checks the mean and standard deviation of the variable. Second, it generates a simulated variable that follows a normal distribution but has the same mean and standard deviation as the observed variable. Third, it orders the simulated variable and the observed variable from their lowest to highest values. Fourth, it produces a scatter plot with the simulated variable on one axis and the observed variable on the other axis. If the observed variable is normally distributed, the data points line up along a diagonal line drawn by the program. If it is not, the points deviate from a diagonal line. The direction of deviation provides some guidance as to how the observed variable is distributed. Let's first look at some *q-q* plots, and then we will learn how to use one with an actual variable.

Figure 10.6 shows that the observed variable (it is on the x-axis) is skewed slightly, and if we were to investigate further, we would find that it has a longer right tail than a normal distribution. One way to *normalize* such a distribution (this is simply shorthand for saying we are trying to transform it into a normally distributed variable) is to take its square root.

In Figure 10.7, we have a variable that has an even longer right tail than the previous variable. If we observed a histogram of this variable, we would find a few observations at an extreme upper end of the distribution. This type of variable needs a greater transformation effort to pull in the extreme values than is available with the square root function. In this situation, most analysts take the natural logarithm ($\log_e$) of the

Figure 10.6: Hypothetical Quantile–Quantile Plot with Positive Skew

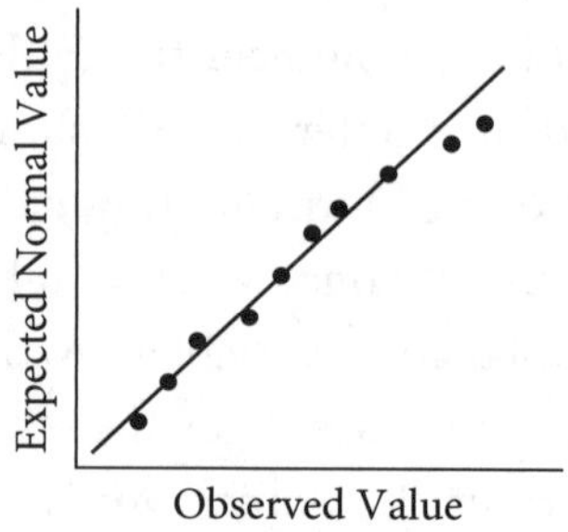

Figure 10.7: Hypothetical Quantile–Quantile Plot with High Positive Skew

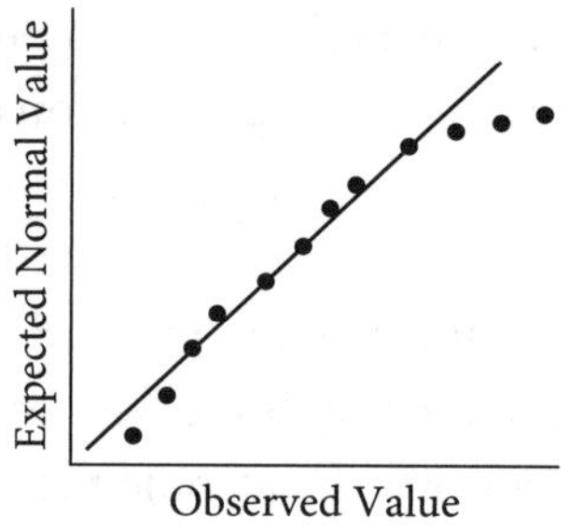

variable. However, if you do this, make sure you recode any zero values to a positive number (perhaps by adding a one to the variable) since the natural logarithm of zero is undefined. A regression model with a logged outcome variable is so popular in applied statistics that it is referred to as a *log-linear* model.

The distribution of the variable in Figure 10.8 is virtually the mirror image of the first *q-q* plot we saw. Its right tail is shorter than a normal distribution, and it has fewer observations in its upper tail than does a normal distribution. To normalize it, we might square its values to stretch them out a bit. Finally, let's see an even more extreme example of this short-tailed distribution (Figure 10.9). In this situation, one might try exponentiating the variable (using the exponential and raising it to the power of the observed value, e^{xi}) to stretch out its distribution.

Of course, this does not exhaust the ways we might transform variables so that they are normally distributed. Moreover, it is important to realize that these transformed variables may be placed in a linear regression model just like any other outcome variable, but the interpretations of the coefficients must change to account for the transformation. For example, if we estimate a model with the square root of income as the outcome variable and age as the explanatory variable, the slope indicates differences in *square root units* of income that are associated with each one-year difference in age.

Figure 10.8: Hypothetical Quantile–Quantile Plot with High Positive Skew

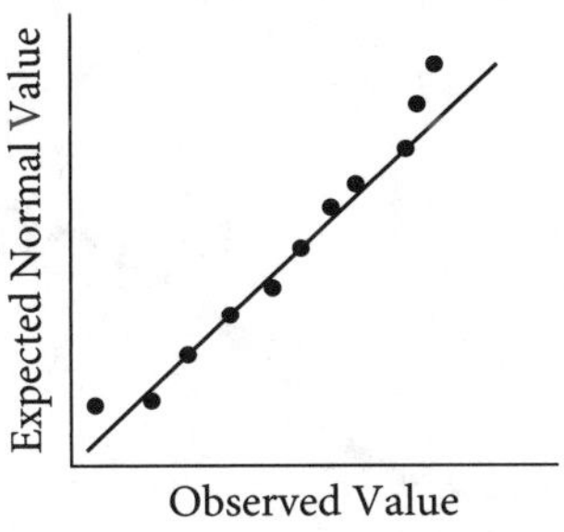

Figure 10.9: Hypothetical Quantile–Quantile Plot with Substantial Positive Skew

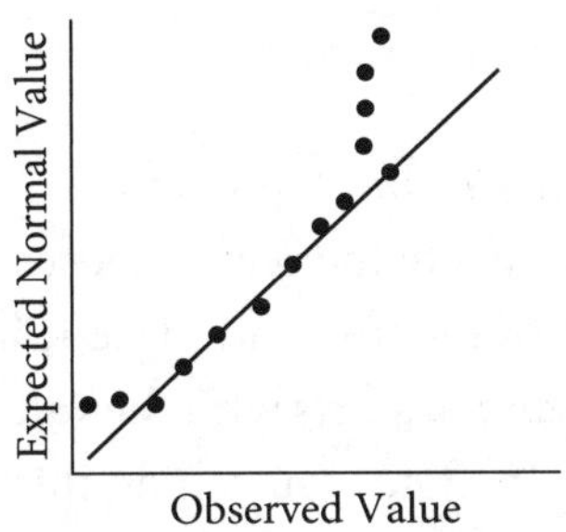

Thus far, we have discussed what we should do to test whether the outcome variable is normally distributed before estimating a linear regression model. However, it is actually more important that we examine the residuals after estimating the model to make sure they follow a normal distribution. After all, this is the key assumption that needs to be examined. We still use the normal *q-q* plot. It is constructed using the `qnorm` command. To see one of these in action, let's estimate a model with the GSS 2010 data set. Once again, we will use personal income as the outcome variable.

First, create a normal *q-q* plot using the variable *rincom06* (in Stata: `qnorm rincom06`; in SPSS, you can use the drop-down menu *Graphs*-Q-Q). An example of the graph you get can be found in Figure 10.10.

Given this, what type of transformation should be used to normalize the income variable? (Note what each axis represents.) Before deciding, let's run a linear regression model with income as the outcome variable, and with age and education as explanatory variables. Next, predict the standardized values (*z* scores) of the residuals (in Stata, `predict rstand, rstand`; in SPSS, while setting up the model, ask it to produce a normal probability plot by checking the box that appears in the Plots window). Now, let's look at a *q-q* plot of these standardized residuals.

Figure 10.10: Quantile–Quantile Plot Produced by Stata

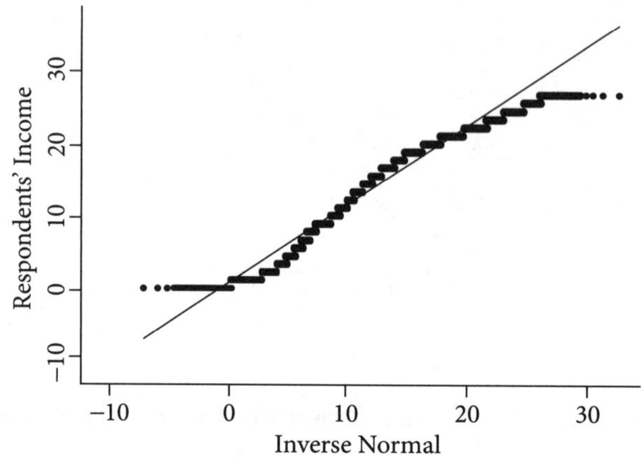

Notice that the residuals of *rincom06* are somewhat off the line, particularly at higher values of income. Can we make them even closer to a normal distribution? Let's compute the natural logarithm of income and place it in the linear regression model. Note, however, that because *rincom06* starts with the value of one, we do not need to use a positive start for this variable. In Stata, you can do this by using the following syntax: `generate ln_inc= ln(rincom06)`. In SPSS, you can use the Transform-Compute drop-down menu and look for the LN function to create this variable; or if you prefer, you can use the syntax file in SPSS by typing the following commands:

```
COMPUTE ln_inc= LN(rincom06).
EXECUTE.
```

Note that the term *ln_inc* is an arbitrary, yet properly descriptive, name. Once we have created this transformed variable, substitute it for the original outcome variable and reestimate the model. What does the normal probability plot show?

It appears we have done more harm than good. There are now a lot of residuals in the middle of the distribution that are well above the diagonal line and well below the line in both the high and low tails. Go back and see whether the square root of income does a better job. What do you find? It looks better than the logarithmic version, although there are still some odd-looking residuals. Is the normal probability plot from the original model better looking? It appears to be. Therefore, even though the outcome variable is slightly skewed, it is suitable for a linear regression model with these two explanatory variables.

Let's now look at a variable that is highly skewed: number of children (*childs*). A *q-q* plot shows the degree of nonnormality. Estimate a simple linear regression model with *childs* as the outcome variable and education (*educ*) as the explanatory variable. What does the normal probability plot show? It looks rather odd, with some residuals well below and some well above the diagonal. Try out a few transformations for number of children to see whether you can normalize the residuals. Then compare the regression model before transforming the variable to one after transforming the variable. Do you see any differences that would make you prefer one model over the other? Both the square root and natural logarithm does a pretty good job of normalizing the residuals, but can you do better? A key question is whether your interpretation of the association between children and education changes after transforming the variable. To put another spin on it: Check the distribution of education. Is it normally distributed? Can you find a transformation that normalizes it? Once you find this transformation, try rerunning the linear regression model with normalized versions of both variables. What is your interpretation of the association now? We think that the best transformation is a logged childs variable and square-rooted education variable or a square-rooted childs variable and a square-rooted education variable.

It is important to keep in mind that you will rarely find a variable in the social and behavioral sciences that has residuals that fall directly on the diagonal line of a normal *q-q* plot. The goal is to find a transformation, assuming one is needed, which allows the observations or residuals to come close to the diagonal line. Scanning the functions in Stata's help menu will give you a good idea about the many transformations that are available. (For advice on how to use these various transformations and numerous other ways to think visually about the distribution of variables, an excellent resource is Cleveland, 1993.)

To summarize what we have learned so far: It is important to check the distributional properties of the variables in the model. For example, we have now learned that it may be important to examine the distributions of the explanatory variables. In fact, some statisticians recommend that analysts should first check to make sure the explanatory variables are normally distributed (with the exception of dummy variables; see chapter 6). Moreover, if the association between an explanatory and outcome variable is nonlinear, it may be best to transform the explanatory variable in some way so that the association is linear or so that it can accommodate a nonlinear association. We saw one example of this when we analyzed the association between age and personal income, but there are many other nonlinear associations to consider. Perhaps taking the natural log or cube root of the explanatory variable is needed to *linearize* (a sophisticated way of saying "make linear") its association with the outcome variable. An association may be "piecemeal" linear, with different slopes needed in different areas of the explanatory or outcome variable's distribution, or we may need to take the squared values of one variable and the cubed values of the other to linearize the association. It is easy to become overwhelmed quickly with the many possibilities, so it is best to be cautious, yet thorough, and always keep in mind that theory and past research usually provide good guidance.

After we have checked the distributions of the variables and assessed whether there are nonlinear associations among the explanatory and outcome variables, it is always a good idea to examine the partial residual plots for each explanatory variable. Sometimes, nonlinear associations emerge in multiple linear regression models that bivariate scatter plots fail to reveal. This has to do with statistical control in a way that is beyond the scope of this presentation. Nevertheless, these nonlinearities may reveal unexpected, yet important, aspects of the model. If any of these nonlinear associations emerge, you should go back and think about why they occur. Including higher-order terms in a model (such as age^2 and age^3) may be interesting and important for correct model specification (see chapter 7 and the age–income example in the current chapter), but it is an empty exercise in the absence of thoughtful consideration about the theory and previous research that initially motivated your conceptual model.

Most important, it is essential that linear regression models be accompanied by normal *q-q* plots of the residuals. A key assumption is that the error term is normally distributed, so testing this assumption is a vital exercise. If you find that the residuals do not follow a normal distribution, there are myriad transformations available. With additional experience looking at normal *q-q* plots, you will find that selecting an appropriate transformation will become easy.

Interaction Terms in Linear Regression Models

This section discusses an important topic in regression analysis: the use of *interaction terms* to represent associations between an explanatory and outcome variable that vary based on a third variable. These are also known as *nonadditive* associations because, to include an interaction term in a linear regression model, we must multiply variables. Therefore, our regression equation is no longer limited to plus signs between variables. For instance, we have, thus far, restricted regression equations to the following form:

$$y_i = \alpha + \beta_1 x_{1i} + \beta_2 x_{2i} + \beta_3 x_{3i} + \varepsilon_i$$

However, there is no reason why we should limit the model so that the predictor variables are completely independent, which is implied by the additive terms in the equation. An interaction term—also known as a *multiplicative term*—involves multiplying explanatory variables. Here is a sample regression equation that we will learn to use shortly:

$$y_i = \alpha + \beta_1 x_{1i} + \beta_2 x_{2i} + \beta_3 (x_{1i} \times x_{2i}) + \varepsilon_i$$

The last term in this equation is listed as $x_{1i} \times x_{2i}$, which indicates that the variables are multiplied. We typically say that they "interact" in some fashion. (A good overview can be found in Aiken and West, 1991.)

Figure 10.11 Visual Example of Dichotomous Variable

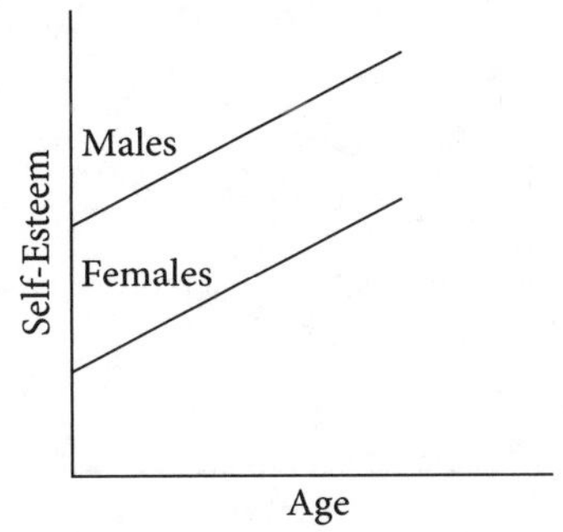

A useful way of thinking about interaction terms is by considering distinct slopes for the different groups represented in the sample. Suppose, for example, that we hypothesize that self-esteem increases between the teenage years and young adulthood (for example, from ages 15 to 25). For now, we will assume that there is a linear association between age and self-esteem within this age range. Given the types of regression equations we have used thus far, we might estimate the following linear regression model:

$$\text{Self-esteem}_i = \alpha + \beta_1(\text{age}_i) + \beta_2(\text{gender}_i) + \varepsilon_i$$

In this model, we treat age as a continuous variable and gender as a dummy variable. Let's say that gender is coded as 0 = female and 1 = male. Suppose our regression output indicates positive slopes for both age and gender, so that males, on average, report higher self-esteem than females. The assumed slopes for age by the two gender groups may be represented using Figure 10.11. If this is confusing, trying making up a couple of slopes for age and gender and then estimate a few predicted scores by using the linear regression equation. You'll notice that the relative distance between the males and females at each age is the same. This type of model is known as a *different intercept–parallel slopes* model. The slopes are the same for the two groups; they are just a constant distance apart. This distance is gauged by the intercept.

Perhaps you've read the literature on self-esteem and suspect that self-esteem increases more among males than among females from ages 15 to 25. Thus, we need a way to model different slopes for males and females. One approach we might use is to estimate separate linear regression models for males and females and then compare the slopes from each model. This is relatively inefficient, however, and does not tell us whether the difference in the slopes is statistically significant. As we'll learn, interaction terms are also very handy for modeling different slopes for two or more groups. Figure 10.12 shows how this is represented, but how do we characterize this type of

Figure 10.12: Example of Continuous–Dichotomous Interaction

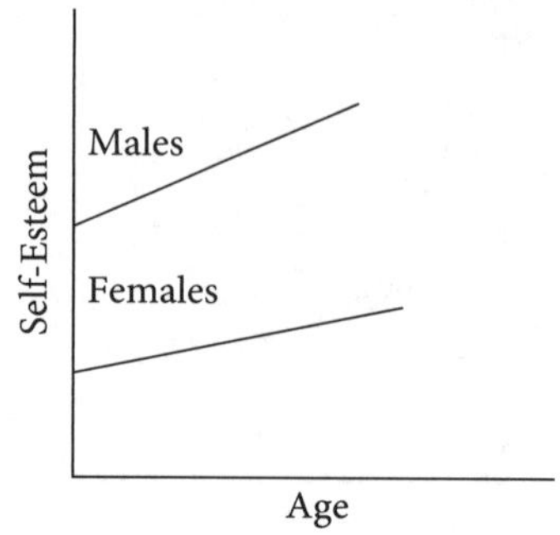

association in a linear regression model? With an interaction term that multiplies age and gender. In the regression equation, this appears as

$$\text{Self-esteem}_i = \alpha + \beta_1(\text{age}_i) + \beta_2(\text{gender}_i) + \beta_3(\text{age}_i \times \text{gender}_i) + \varepsilon_i$$

As we saw earlier, the values of one explanatory variable are multiplied by the values of the other explanatory variable; in this case, age is multiplied by gender. If the coefficient for β_3 is statistically significant, then we may infer that males and females have different age–self-esteem slopes. Hence, this type of model is known as a *different intercept–different slopes* model. Normally, though, we need to consider the other coefficients as well. At this point, we are assuming that all three are positive and statistically significant. This also raises another important issue when using interaction terms: If interaction terms are used in the model, the constituent variables must also appear in the model for correct specification.

Let's see an example of an interaction term using the GSS 2010 data set. We will consider the associations among age, personal income, and gender, but first, consider the following issue: We know that education and income are positively associated in the United States, with people who have more years of formal education, on average, earning higher incomes than people with fewer years of formal education (an interesting exception involves air traffic controllers, but we will ignore this). Many studies also indicate that women, on average, earn less than men. Let's see whether these two observations are supported in the GSS data set. We will include age as a control variable in the model because we know it is associated with both education and income (but, to simplify things, we will ignore the nonlinear association between age and income).

The results shown in Table 10.3 support both propositions: Controlling for the effects of age, males, on average, report more personal income than females; and more years of education is associated with more personal income (note that *gender* is coded as 0 = female and 1 = male in this data set). If we were to plot slopes for education,

Table 10.3: Personal Income Regressed on Education and Gender

Variable	*b*	*SE*
educ	0.810	0.055***
gender	2.682	0.327***
Intercept	1.551	0.814
N	1,200	
R^2	.183	

***$p < .001$.

they would appear similar to Figures 10.11 and 10.12 shown earlier: different intercepts but parallel slopes for males and females. The average difference between males and females is 2.682 units of income.

Now, let's consider another question that involves these variables: What role does gender play, if any, in the association between education and income? Does more education tend to equalize personal income among males and females, or is there some type of glass ceiling that results in a greater income disparity among males and females at higher levels of education? Perhaps the occupational choices of highly educated men and women lead to an income disparity. There may be no association between gender and education when it comes to personal income; they may only be independently associated with this outcome. We cannot even approach the answers to these questions with the linear regression model we just estimated. It provides no evidence about whether income varies by gender *and* education. Fortunately, an interaction term may be used to gather some of this evidence.

In Stata, we may include the interaction term using the following regression command: `regress rincom06 age c.educ c.gender c.educ#c.gender` (an alternative is to create the interaction term using the `generate` command). In SPSS, you can use the *Transform-Compute* drop-down menu to calculate the interaction term. Alternatively, in the SPSS syntax, file you can type:

```
COMPUTE gen_educ = gender * educ.
EXECUTE.
```

The results can be found in Table 10.4.

How should we interpret these results? First, by comparing them to the results of the previous model, we see that the education coefficient and the gender coefficient are larger (in absolute terms). Second, the Gender × Education interaction term has a negative and statistically significant coefficient. It takes some experience to be able to look at the three slopes—education, gender, and Gender × Education—and make sense of the patterns, but here is a hint that will help you interpret them: Choose one of

Table 10.4: Personal Income Regressed on Education and Gender, with Interaction

Variable	*b*	*SE*
educ	1.000	0.080***
gender	8.161	1.553***
educ*gender	−0.399	0.109***
Intercept	−4.119	1.230
N	1,200	
R^2	.217	

***$p < .001$.

the variables involved in the interaction to focus on and then consider what the interaction term is telling you about this slope. Let's pick education. Its slope is 1.000, and it is significantly different from zero, but the interaction term multiplies either a zero (to represent females) or a one (to represent males) by education. Therefore, for males, we assume one slope, but when we consider females, the slope is pulled in a downward direction (as indicated by the negative gender × education slope). Similarly, females begin with a detriment in income (as shown by the positive slope for gender) but then this gap diminishes at higher levels of education (as indicated by the negative gender × education slope).

If this is still difficult to understand, you can use either Stata or SPSS to compute predicted values for some representative groups and then compare these values to see whether the gap between males and females changes at different levels of education.

Stata

A simple way to compute predicted values is with Stata's margins command. Here is an example:

```
regress rincom06 age i.educ i.gender educ#gender
margins educ#gender
```

Noting that gender is coded as 0 = female and 1 = male, Stata provides some expected values based on one's education level and gender. For example, at 12 years of education, men are expected to have an income of 14.35 units, while women have an expected income of 11.01 units, a difference of 3.34 units. At 20 years of education, the advantage for men has shrunk to 0.68 units. Thus, it appears that the gap between males and females on personal income is greater at lower levels of education.

SPSS

SPSS can also give you predicted values for some representative groups and then compare these values to see whether the gap between males and females changes at different levels of education. To do this, rerun the regression model and ask SPSS to save the unstandardized predicted values. Then use the *Analyze-Compare Means-Means* drop-down menu to examine the means for the different gender and education groups represented in the data set. Place the unstandardized predicted values (these are labeled *PRE_1* by SPSS) in the box labeled Dependent List and *educ* in the box labeled Independent List. Then click the Next box under Layer 1 of 1. This allows you to place another layer in the Means command. Place *gender* in the Independent List box and click *OK*. Here, we find the same results that we found using Stata.

Another way to think about the difference is by figuring the percentage difference between men and women at different levels of education. To do this, use the following simple calculation:

$$\text{Percent difference, 12 years: } \frac{\{14.35 - 11.01\}}{11.01} \times 100 = 30.3\%$$

$$\text{Percent difference, 16 years: } \frac{\{18.38 - 17.70\}}{17.70} \times 100 = 3.8\%$$

It is now clear that there is a larger percentage difference between males and females with 12 years of formal education than with 16 years of formal education. This supports the idea that higher levels of education tend to diminish income differences among men and women in the United States.

However, you may have anticipated a potential problem that could affect our confidence in the regression model. Recall that an interaction term is calculated by multiplying one variable by another. This is bound to lead to collinearity between the interaction term and its constituent terms. In fact, if we rerun the linear regression model and ask Stata for collinearity diagnostics, we find that *gender* has a VIF of 23.6 and the interaction has a VIF of 24.15. Earlier in the chapter, when computing higher-order terms for age, we learned that a solution to this problem is to standardize (that is, take the *z* scores of) the constituent variables first before computing the higher-order terms. It makes no sense to standardize gender as it is a dummy variable, so let's standardize *educate* and then compute the interaction term. Then place age, gender, the standardized value of education, and the interaction term in the linear regression model designed to predict income. The results of this new model show a lack of multicollinearity but the same general patterns that we saw earlier. In fact, we may get predicted values in SPSS or Stata as we did earlier and find identical results.

Interaction Terms with Continuous Variables

It is slightly more difficult when we wish to use interactions among two continuous variables, although the same logic and most of the same procedures apply. First, we need to have a reason to include interaction terms: Why do we think they belong in a regression model? Second, we may multiply the two continuous explanatory variables and place the interaction term and its constituent terms in the regression model. There is bound to be multicollinearity, so you may wish to calculate *z* scores for the two variables and use them to compute the interaction term.

Let's look at an example of an interaction term that uses two continuous variables. We will return to the GSS 2010 data set and consider some predictors for number of children a respondent has. In this example, we suspect that education (*educ*) and number of siblings (*sibs*) have an effect on number of children, but we also think that these two variables may work together to influence number of children. We are unsure whether this prediction is valid, so we will use a two-tailed hypothesis test and simply propose that education and siblings interact in some way.

To test this notion, let's estimate a linear regression model. First, though, because we wish to minimize the risk of multicollinearity, we will use standardized versions of the explanatory variables (*educ* and *sibs*) and calculate the interaction term between them by using the standardized forms of these variables. After calculating the interaction term, we may estimate the regression model. The results are provided in Table 10.5.

Education, as expected, is negatively associated with number of children. Because it is measured in *z* scores, we may interpret its slope as "Controlling for the effects of the other variables in the model, each one standard deviation increase in education is associated with 0.454 fewer children." Further, we found that number of siblings is positively associated with number of children. However, what does the coefficient of the interaction term show?

One way to think about the interaction term is to consider the slope for education, which is negative. However, as number of siblings increases, this negative slope becomes even more negative. Another way to view it is to consider the slope of siblings, which is positive. As education increases, this positive slope moves in a negative direction (it is less steep).

Although such interpretations provide general guidance for understanding the interaction term, some analysts find it more intuitive to compute predicted scores. The means comparison procedure mentioned above is not very useful unless you wish to create categorical variables from the continuous variables, so we will simply compute the predicted values by hand (and calculator), but we have to decide on the values of the explanatory variables. Because they are all measured using *z* scores, convenient values include −1, 0, and 1; or one standard deviation below the mean (low), at the

Table 10.5: Number of Children on Education and Number of Siblings

Variable	*b*	*SE*
zeduc	−0.454	0.038***
zsibs	0.194	0.046***
zsibs*zeduc	−0.197	0.046***
Intercept	1.864	0.037
N	2,037	
R^2	.095	

***$p < .001$.

Table 10.6: Predicted Number of Children, by Education and Number of Siblings

	Education		
Siblings	**One *SD* below**	**Mean**	**One *SD* above**
One *SD* below	1.917	1.660	1.403
Mean	2.307	1.853	1.399
One *SD* above	2.697	2.046	1.395

mean (medium), and one standard deviation above the mean (high). Let's see what the predicted number of children is at each of these values. We may use the following equation and plug in the values for the siblings and education:

$$\text{Children} = 1.853 + .193(\text{zsibs}) - .454(\text{zeduc}) - .197(\text{zsibs} \times \text{zeduc})$$

What do these predicted values in Table 10.6 indicate about the association among siblings, education, and number of children? Notice, first, that the lowest predicted number occurs when an individual has a high number of siblings and high education. The highest predicted number occurs when someone has low education and high number of siblings, but how can we determine the effects of, say, siblings on the association between education and the number of children? We may compare the predicted values down the columns. For instance, at low values of percent education, the predicted values increase by about 20 percent, whereas at high values of education, the predicted values decrease by about 1 percent. In other words, the slope of the education changes with number of siblings. You could undertake a similar process focusing on the influence of education on the relationship between siblings and children by comparing across the rows.

As a final note, remember to check the normal *q-q* plot of the residuals from this model. Sometimes interaction terms introduce additional complications in a model, such as nonnormally distributed residuals. What does this model show? Are you more or less confident in the results after examining the distribution of the residuals? More important, though, what does the literature on fertility tell us about the model? Do we trust the finding that the interaction should be negative?

Before concluding this chapter, we should mention that regression models are not limited to interaction terms involving only two explanatory variables, or what are known as *two-way* interactions. You may also enter three-way interaction terms in a regression model, which implies multiplying three explanatory variables together (for example, $x_1 \times x_2 \times x_3$). However, this regression model must include all of the constituent terms as well as all the possible two-way interactions ($x_1 \times x_2$; $x_1 \times x_3$; $x_2 \times x_3$). Conceivably, four-way, five-way, or even higher-order interactions are possible, but these quickly become very difficult to work with and are even harder to interpret.

We close the chapter with a final word of warning. There are some statisticians who claim that interaction terms are problematic because they fail to disentangle the actual ordering of the variables. Recall in chapter 7 that we discussed endogeneity or whether there are explanatory variables in the regression model that are predicted by (or, to use generally unrecommended terminology, caused by) other explanatory variables inside or outside the model. If one of the variables used to compute the interaction term is endogenous to the other—or to other variables—then the usual interpretation of the interaction term becomes questionable. Think about the model estimated earlier in the chapter: Perhaps gender in some way influences education, or, as we say, education is endogenous in the model. What about the fertility model? Number of siblings may affect education (for example, the more siblings you have, the less educated you are), so education could be endogenous in the model. A comprehensive discussion of this problem is beyond the scope of this presentation, but you should always think carefully about the variables in your model when considering interaction terms and the likelihood of endogeneity.

Some Suggestions

This chapter has covered a lot of important material. Nonlinear associations and outcome variables—as well as residuals from linear regression models—that are not normally distributed are quite common in the world of regression analysis. Moreover, it is important to consider that many of the associations between two variables that interest you may depend on some third variable. Interaction terms provide a valuable way to test for nonadditivity in your models. As a review and conclusion to what we have learned, here are four suggestions:

1. Always use theory and previous literature to guide your thinking about interaction terms and possible nonlinearities among your variables.
2. Construct bivariate scatter plots of each explanatory variable by the outcome variable. Plot nonlinear lines in these scatter plots. Check the partial residual plots for the regression model by using standardized or studentized residuals. Check the normal *q*-*q* plot of the residuals. When faced with nonlinear associations, consider the various transformations that are available. It is likely you will find one that will linearize the associations or normalize the residuals.
3. When entering interaction terms, consider using standardized versions of the explanatory variables to diminish the risk of multicollinearity. Use predicted values to figure out what the interaction term implies about the associations in the model.
4. It is easy to get carried away by combining quadratic terms, cubic terms, interactions, and transformed variables in a single regression model. Take care to avoid doing too much because models can easily become unmanageable when there are too many of these types of variables. It is better to have a good idea—guided by theory and previous research—of whether these associations exist, rather than searching for them using multiple models. Remember that one goal of good statistical models is parsimony. If you cannot explain them, then perhaps they should not appear in your model.

Key Terms

The following is a list of key terms and concepts in chapter 10. You can find definitions to the words throughout the chapter.

- Ceiling effect
- Different intercept–different slopes model
- Different intercept–parallel slopes model
- Floor effect
- Higher-order terms
- Interaction terms
- Multiplicative terms
- Nonadditive effects
- Nonlinear
- Nonparametric regression
- Normal quantile–quantile or normal *q*-*q* plot

Analysis Exercises

1. Generate a variable for the natural log of siblings (*sibs*), a variable for the natural log of income (*rincom06*), and a variable for the natural log of educational attainment (*educ*). Remember that the log of 0 cannot be defined. Thus, you will need a "positive start" for any variable with a true zero. Run the following data analyses:
   a. Number of children as a function of log of siblings
   b. Logged income predicted by educational attainment
   c. Logged income predicted by logged educational attainment
2. Generate a variable for the square of age. Is the relationship between number of children and age curvilinear in nature?
3. Run a model that predicts mental health well-being (*mntlhlth*) as a function of education (*educ*), physical health (*physhlth*), and the interaction of education and physical health. What do the results show you?

Solutions

1. a. A 1 percent increase in siblings increase the number of children by approximately 0.006 children.
   b. A one-year increase in education is associated with a 7 percent increase in income.
   c. A 1 percent increase in education is associated with a 0.81 percent increase in income.
2. Yes, the effect of age on number of children increases at a decreasing rate.
3. The effect of education on poor mental health is negative (–.162), while poor physical health is shown to lead to poorer mental health (.551). The interaction (–.02) shows that as education increases, the effect of poor physical health becomes less positive, while poor physical health makes the effect of education slightly more negative.

CHAPTER 11 HETEROSCEDASTICITY AND AUTOCORRELATION

Chapter Objectives

This chapter discusses the causes of, consequences of, and solutions to two common issues in OLS regression: heteroscedasticity and autocorrelation.

In chapter 11, we will discuss

- what heteroscedasticity is and why it is problematic
- how to test and correct for heteroscedasticity
- the sources of autocorrelation, what it is, and why it can be a problem
- tests and corrections for autocorrelation.

An important assumption of the linear regression model is that the variance of the error term is constant for all combinations of *x*s. When we collect information from a sample of adults in the United States, we assume that each adult represents some number of adults in the population of U.S. residents. In an important way, each adult's value on some variable of interest represents the mean value of the variable in the subpopulation he or she represents. Let's take it for granted that these sample members' values are a good representation of the means for the subpopulations. What about variability? Is it the same across sample members who have different values for some set of explanatory variables?

To drill a bit deeper, consider the model in chapter 4 designed to predict days of poor mental health in the past month. In the example, we used the following variables to predict the outcome variable: days of poor physical health, years employed, and income. We now know that we made several assumptions when estimating this linear regression model (for example, lack of measurement error in the variables; correctly specified model). Let's focus on just one explanatory variable—days of poor physical health—to understand one of the assumptions addressed in this chapter. We assumed that the variability in the errors in predicting poor mental health followed the same pattern for any particular value of physical health. This means that we assumed the same variability for these errors whether people had no days of poor physical health or 30 days.

Figure 11.1: Heteroscedastic Errors

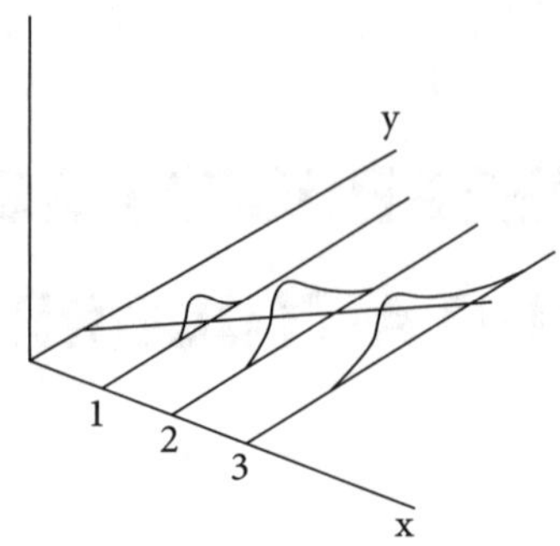

The name for this assumption is *homoscedasticity*. The prefix *homo-* means *same*, and the term *scedastic* means *scatter*. Hence the errors are assumed to be homoscedastic or have the same scatter. The alternative is that the errors do not follow the same pattern; hence, they are *heteroscedastic* (*hetero-* means different). Figure 11.1 shows an example of heteroscedastic errors. Note how the presumed spread of the errors in predicting y increases with hypothetical values of x.

The main consequence of heteroscedastic errors is that, although the slopes are unbiased, the standard errors are biased. An alternative way of saying this is that although the slopes are unbiased, they are *inefficient*. The degree of bias is rarely known, although it seems to be especially severe in small samples. Yet it can throw off interpretations of statistical significance even in moderately sized samples.

Heteroscedastic errors are a common phenomenon in linear regression models. They seem to be particularly common in cross-sectional data, or data that were collected at one point in time. We will return to the GSS 2010 data to look at the relationship between education and mental health. A general hypothesis is that more educated individuals have better mental health than less educated people. The results of a simple linear regression analysis, seen in Table 11.1, support this proposition. Each one-year increase in education is associated with about a 0.275-day decrease in poor mental health.

One of the graphical tests we should consider after estimating this model is a partial residual plot. In chapter 10, we learned that this type of plot is useful for checking the appropriateness of the linear model. However, it also provides a useful way to check whether the errors are homoscedastic or heteroscedastic in a simple linear regression model. You can ask either Stata or SPSS to construct a simple scatter plot of the studentized residuals (y) by the income variable (x). Figure 11.2 displays this graph.

Although it is a little hard to tell with so many values, it appears that the errors are not randomly distributed. A graph of heteroscedasticity could show the errors fanning out at higher values of the explanatory variable or the residuals narrowing down at

Table 11.1: Poor Mental Health Regressed on Education

Variable	*b*	*SE*
Education	−0.275	0.071***
Intercept	7.674	1.010***
N	1,149	
R^2	.013	

***$p < .001$.

Figure 11.2: Stata Residual Plot to Diagnose Heteroscedasticity

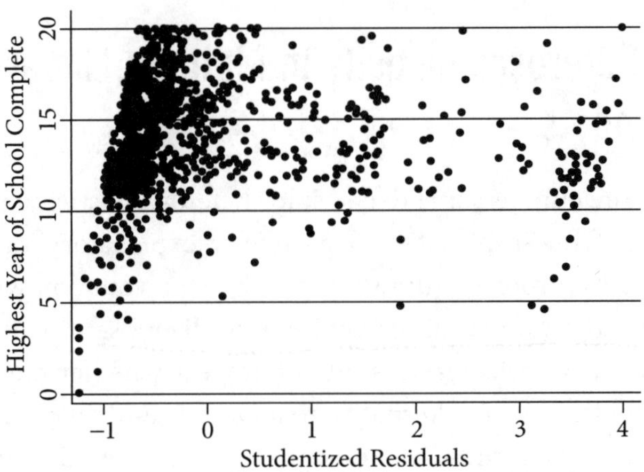

higher values of the explanatory variable. In contrast, homoscedastic residuals look like a random scatter, with no recognizable association in the partial residual plot.

At this point, it is important to ask yourself the "why" question: Why does this pattern exist? Is there a reasonable explanation? Often, by answering this question in a thoughtful way, you can figure out a solution to the problem of heteroscedasticity. For instance, it seems reasonable to suggest that different kinds of mental health issues might be associated with different levels of education or that mental health insurance coverage varies by education level. How does this offer a solution to heteroscedasticity, though? If we had a variable in the data set that measured insurance coverage, then we could include it in the model and determine whether it explained the heteroscedasticity among the residuals. This illustrates that, as with many problems that affect the linear regression model, a good theory goes a long way.

Another common reason for heteroscedasticity is that one or more of the variables requires transformation. This is often not apparent by looking at a scatter plot between the outcome and one of the explanatory variables, but previous research may provide a clue. As mentioned in an earlier chapter, income measures are often linearized by taking their natural logarithms, and many statisticians point out that taking the natural logarithm of the explanatory or outcome variable may be all that is needed to eliminate the heteroscedasticity problem. For example, take the natural logarithm of either the explanatory or outcome variable in the linear regression model we just estimated, reestimate the model using the transformed variable, and construct a partial residual plot. What does the plot show? There does not appear to be a clear heteroscedastic pattern to the residuals, although it does look unusual (either like a U-shaped or inverted U-shaped pattern, which may indicate other problems that we don't have the space to investigate).

Checking for Heteroscedasticity in Multiple Linear Regression Models

It is only slightly more cumbersome to check for heteroscedasticity in a multiple linear regression model. If you suspect that a particular explanatory variable is involved in a heteroscedasticity problem (draw on theory and your knowledge of previous research!), then construct a partial residual plot to diagnose it. However, when there are multiple explanatory variables, it is often not clear whether one or more of them may induce this problem. As an alternative to a partial residual plot, most statisticians recommend plotting some form of the residuals against the predicted values. If you are a Stata user, you can automate this process with the regression postcommand `rvfplot` (residuals-vs.-fitted values plot), which provides a scatter plot of the unstandardized residuals by the predicted values. However, John Fox (1997), in his books on regression modeling, recommends plotting the studentized residuals (or deleted studentized residuals) (y) against the standardized predicted values (x). This is because the studentized residuals tend to have more constant variance than the standardized residuals. Regardless of the type of residual used, we still look for a fan-shaped pattern in the scatter plot. Of course, we prefer to find a random pattern because this suggests that the errors are homoscedastic.

However, we have determined that in real data sets, such as the GSS 2010, it is sometimes hard to find heteroscedasticity graphically. Thus, we might ask, are there other ways to detect it? One of the problems is that graphical techniques become difficult to use with large data sets (imagine trying to visualize patterns with hundreds of points in a scatter plot), so it would be nice to have some other techniques for diagnosing heteroscedasticity. Fortunately, there are several. We will discuss two of these.

The first diagnostic approach is known as *White's test* (after the econometrician H. White). This test is conducted after estimating a multiple linear regression model. It is composed of the following steps:

1. Compute the unstandardized residuals from the regression model.
2. Compute the square of these values (for example, resid^2).
3. Compute the square of each explanatory variable (for example, x_1^2, x_2^2). Do not compute squared values of dummy variables, though.
4. Compute two-way interactions using all the explanatory variables (for example, $x_1 \times x_2$; $x_1 \times x_3$; $x_2 \times x_3$).
5. Estimate a new regression model with the squared values of the residuals as the outcome variable and the computed variables in steps 3 and 4 as the explanatory variables (don't forget to include their constituent terms).
6. There are then two ways to conclude whether or not the errors are heteroscedastic: (a) Rely on the R^2 from the model in step 5; if it is significantly different from zero, then there is heteroscedasticity. (b) Use the following test statistic:

$$nR^2 \sim \chi^{2_k},$$

where n = sample size, R^2 is from the model, and k = the number of explanatory variables. Note that this test statistic is distributed as a chi-square variable, so we need to compare its value with a chi-square value with k degrees of freedom.

An advantage of White's test is that not only does it test for heteroscedasticity, but it also provides evidence about potential nonlinearities in the associations. Unfortunately, SPSS does not automatically calculate White's test. However, Stata will save us the trouble of having to go through each of these steps. Although slightly different, the postcommand `estat imtest, white` will produce several tests, including White's test for heteroscedasticity. The White's test for the model in Table 11.1 is shown in Figure 11.3.

The first part of the printout in Figure 11.3 shows that White's test compares the null hypothesis of homoscedasticity against the alternative of heteroscedasticity. We should reject the null hypothesis in this situation (note the small p value) and conclude that there are heteroscedastic errors in this model.

Another commonly used technique for detecting heteroscedasticity is a relatively simple one, known as the *Breusch-Pagan test*. In Stata, this test is easy to implement using the postcommand `estat hettest`. The null and alternative hypotheses are the same as in White's test. Again, we reject the null hypothesis and conclude that there are heteroscedastic errors. The Stata output looks like that shown in Figure 11.4.

Figure 11.3: White's Test Output in Stata

```
estat imtest, white

White's test for Ho: homoscedasticity
     against Ha: unrestricted heteroscedasticity

     chi2(2)      =   24.30
     Prob > chi2 =  0.0000

Cameron & Trivedi's decomposition of IM-test
```

Source	chi2	df	p
Heteroskedasticity	24.30	2	0.0000
Skewness	138.33	1	0.0000
Kurtosis	89.18	1	0.0000
Total	251.81	4	0.0000

Figure 11.4: Breusch-Pagan Test in Stata

```
estat hettest

Breusch-Pagan / Cook-Wisberg test for heteroskedasticity
          Ho : Constant variance
          Variables : fitted values of mntlhlth

          chi2(1)       =  84.35
          Prob > chi2   =  0.0000
```

In SPSS, the process is a little more involved but can be achieved by doing the following:

1. Run the regression and save the unstandardized residuals and unstandardized predicted values.
2. Create a new variable that is the square of the residuals.
3. Get the sum of the squared residuals by using the drop-down menu *Analyze-Descriptive Statistics-Descriptives* and selecting *Sum*.
4. In the output viewer, copy the value in the sum cell of the descriptive statistics output.
5. Compute a new variable with the following formula: [squared residual/(sum of squared residuals/total observations)].
6. Estimate a linear regression with this new variable as the dependent variable and the predicted values as the independent variable. Divide the regression sum of squares in the ANOVA table by 2 and find the associated p value for this number in an ANOVA table.

Solutions for Heteroscedasticity

We have already mentioned one possible solution to heteroscedasticity: to figure out why it occurs. What variable or variables might influence it? However, we might have an improperly specified variable and need to transform the variable. For example, transforming a variable by taking its natural logarithm also may reduce heteroscedasticity in the model. This is especially useful when using well-known skewed variables such as income or measures of deviant behavior. Taking the natural logarithm helps when the residuals spread out at higher values of an explanatory variable; if they narrow down, then squaring the variable or exponentiating it may help. Note, though, that these solutions involve the proper specification of the model. Heteroscedasticity is often only a symptom of improper model specification.

However, there are also situations when a variable that you may not be particularly interested in induces heteroscedasticity. This problem emerges quite frequently when we analyze what are known as *repeated cross-sectional data*. The GSS data we have been using are a well-known example of this type of data set. Although we have used only the 2010 data, the GSS has been collected for more than 30 years. A different sample of adults in the United States are surveyed every couple of years; hence, it involves cross-sectional data that are repeated over time.

One of the problems with such data is that the sample size may vary from one survey to the next. Because sample size affects variability (for example, larger samples generally lead to smaller standard errors), differing sample sizes across years—assuming one wishes to analyze data from several years—induces heteroscedasticity. Some analysts simply include a variable that gauges year in their models, hoping that this will minimize heteroscedasticity. A preferred approach, however, is to use *weighted least squares* (WLS) to adjust for differing sample sizes. WLS uses a weight function to adjust the standard errors for heteroscedasticity. The general formulas for WLS are

$$\text{SSE} = \sum \left(\frac{1}{s_i^2} \right) (y_i - \hat{y}_i')^2$$

$$\hat{\beta}_1 = \frac{\sum \frac{1}{s_i^2}(x_i - \bar{x}')(y_i - \bar{y}')}{\sum \frac{1}{s_i^2}(x_i - \bar{x}')^2}, \quad \text{where } \bar{y}' = \frac{\sum \frac{y_i}{s_i^2}}{\sum \frac{1}{s_i^2}} \text{ and } \bar{x}' = \frac{\sum \frac{x_i}{s_i^2}}{\sum \frac{1}{s_i^2}}$$

The formula for the slope is for simple linear regression models, but extending it to multiple linear regression models is not difficult. The key for WLS is the weight function, or $1/s_i^2$. In the example of repeated cross-sectional data, the inverse of the sample size often works well to correct standard errors for heteroscedasticity.

Table 11.2: Weighted Least Squares Regression of Poor Mental Health Days

Variable	*b*	*SE*
Education	−0.277	0.070***
Constant	7.702	1.007***
N	1,149	
R^2	.013	

***$p < .001$.

However, in most situations—especially those that do not involve repeated cross-sectional data—we must come up with the weight function conceptually or by thinking about the associations implied by the model. For instance, suppose that our education–mental health example was not solved by considering nonlinearities. Are there other variables in the data set that might be inducing heteroscedasticity? In the GSS data set, there are many possibilities, so let's propose the following hypothesis: Many individuals get health insurance through their employers, so individuals who work more hours should be more likely to get health insurance. Of course, this is not an ironclad rule. Some people who do not work have good health insurance, and some who work many long hours have no insurance. Hence, when evaluating the association between work and mental health, heteroscedastic errors might be induced by combining individuals who do not work with those who work long hours. If this argument has any merit (and it probably doesn't), then a measure of average hours worked may be just the weight variable we are looking for. In the GSS data set, there is a variable that assesses hours of work in the past week (*hrs1*) that may prove useful.

Now let's see what the WLS regression model with hours does. Stata users have many options available to them, but the most straightforward command is known as `wls0`. This is a user-written command that must be downloaded first (type `findit wls0`). After making sure it is available on your computer, use the following command: `wls0 mntlhlth educ, wvar(hrs1) type(abse)` graph. If you are using SPSS, this regression can be done by using the drop-downs *Analyze-Regression-Weight Estimation*. Although it looks very similar to the linear regression drop-down, it now has a weight variable where you can put *hrs1*. The results of the WLS regression can be found in Table 11.2.

It does not appear as though the WLS model offers much improvement over the linear regression model. Moreover, the accompanying graph still shows heteroscedastic errors. You might try other variations of this model or alternative variables in the data set, although it can quickly turn into a search for the unobtainable. In fact, it is not a good idea to simply hunt willy-nilly for weight variables without strong theoretical justification as using the wrong weight can lead to misleading results.

Figure 11.5: Regression with Robust Standard Errors (Huber-White-Sandwich Correction) in Stata

regress mntlhlth educ, robust

Liner regression

Number of obs	=	1149
F (1, 1147)	=	13.12
Prob > F	=	0.0003
R-squared	=	0.0130
Root MSE	=	7.2756

mntlhlth	Coef.	Robust Std. Err.	t	P>|t|	[95% Conf.	Interval]
educ	−.2747143	.0758398	−3.62	0.000	−.4235147	−.1259139
_cons	7.67381	1.154877	6.64	0.000	5.407902	9.939719

The other solutions to heteroscedastic errors do not require us to find the source of the problem. Rather, they are general solutions. One is named *White's correction* (after the same econometrician who developed White's test) or the *Huber-White-Sandwich estimator* (because it was also described by statisticians Huber and Eicker; we do not know why Eicker isn't recognized), and the other is named the *Newey-West approach* (after the econometricians who developed it). Unfortunately, none of these are available in SPSS directly—although you could search the Internet for user written commands to help you. All involve a lot of matrix manipulation that is beyond scope of this chapter. However, there are Stata commands available that will do these corrections. The simplest approach is to use the `robust` subcommand along with the `regress` command: `regress mntlhlth educ, robust`. Stata returns the output shown in Figure 11.5.

The main difference is in the standard errors. Note that, compared with the original model, the so-called robust standard error is larger than the original standard error. This is usually the case when the errors are heteroscedastic.

Some experts suggest that because heteroscedasticity is such a common problem in linear regression models, we should always use a correction method. If there is no heteroscedasticity, then the results of the corrected and uncorrected models are the same. However, if there is heteroscedasticity, then the results of a standard linear regression analysis can be highly misleading (see Ervin, 2000).

Autocorrelation

Another problem arises when the error terms follow an autocorrelated pattern. This problem is especially acute in two types of data: those that are collected over time and those that are collected across spatial units. Data that are collected over time come in several types. First, we've already learned about repeated cross-sectional data. Second, data that are collected from the same individuals (these can be people, animals, or other individual units) over time are known as *longitudinal* or *panel data*. Third, data that are collected on the same aggregated unit (for example, cities, counties, states) over time are known as *time-series data*. Time-series data are often limited to one unit, such as the city of Detroit, the state of Illinois, or the entire United States; data collected on different aggregate units over time are known as *cross-sectional time-series* data. Regardless of the type of data collected over time, autocorrelation is virtually a constant problem in these data types.

One way to think about this problem is to consider the nature of residuals or errors in prediction. When we collect data over time, there is typically a stronger association among errors in time periods that are closer together than in those that are farther apart. For example, if we collect information on crime rates in New York City over a 25-year period and try to predict them based on unemployment rates over the same period, our errors are likely to be more similar in 1980 and 1981 than in 1980 and 2000. Hence, the errors are correlated differentially depending on time. Another name for this type of autocorrelation is *serial correlation*.

A similar problem occurs when collecting data across spatial units. Suppose we collect data on suicide rates from cities across the United States. We then use percent poverty and the amount of air pollution to predict these rates. The errors in prediction are likely to be more strongly related when considering Los Angeles and San Diego, California, than when considering San Diego and Providence, Rhode Island. Los Angeles and San Diego probably share many more characteristics, some of which we do not measure, than do San Diego and Providence. These unmeasured characteristics add to the error in the model. Another name for this type of problem is *spatial autocorrelation*.

As with heteroscedasticity, the main result of autocorrelation is biased standard errors. The slopes, on average, are still correct, but the standard errors and, ultimately, the t values and p values are not correct. Hence, autocorrelation can make it much more difficult to determine whether a slope is statistically distinct from zero in the population.

The scatter plot in Figure 11.6 shows an example of autocorrelation or, more accurately, serial correlation. Notice the snaking pattern of the residuals around the linear regression line. This is the typical appearance of autocorrelation; it is the consequence of the stronger association among errors closer together in time than among those farther apart in time. Of course, such a pattern is rarely so clear; it is more common, especially with larger sample sizes, to find an unrecognizable pattern to the residuals. Hence, we need additional tools to determine whether autocorrelation is a problem in linear regression models.

Let's see an example of serial correlation. The GSS 2010 data set we've been using does not have this issue, so we will move to a different data set called *Detroit.dta* if you are using Stata or *Detroit.sav* if you are using SPSS. These data include variables that assess

Figure 11.6: Example of Autocorrelation

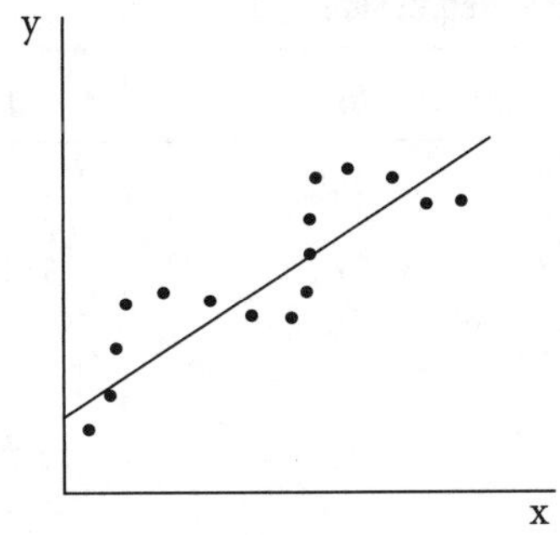

Figure 11.7: Example of Autocorrelation in Stata

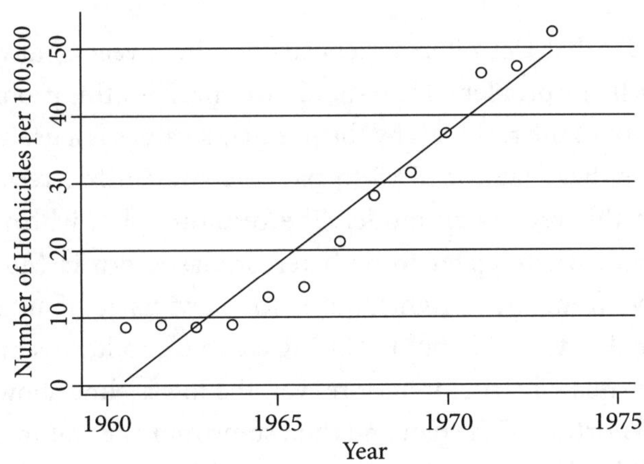

the annual number of homicides in Detroit, Michigan, from 1961 to 1973. The data set also has additional variables that were collected on an annual basis. This is an example of a classic time-series data set. Looking at homicides over time indicates an increasing trend, but let's look at a scatter plot of the homicide rate (y) by year (x) to see whether we can find evidence of autocorrelation. When constructing this scatter plot, use the `lfit` command to include a linear regression line. It should look similar to Figure 11.7.

Suppose, though, that we wish to use one of the variables in the data set to predict the number of homicides. A key issue in criminal justice management is whether the number of police officers affects crime patterns. We will thus consider whether the number of police per 100,000 people predicts the number of homicides. What does a linear regression model show? (We will ignore year for now.) The results on Table 11.3 indicate that the number of police per 100,000 is positively associated with the

Table 11.3: Comparison of Linear Regression and Prais-Winsten Regression

	Linear Regression		Prais-Winsten	
Variable	***b***	***SE***	***b***	***SE***
Police	0.337	0.028***	0.289	0.045***
Constant	−77.630	8.631	−63.373	14.304
N	13		13	
R^2	.929		.662	
Rho	—		0.763	
D-W test	0.666		1.469	

Note: Dash indicates that the statistic was not generated for the specific test.
***$p < .001$.

number of homicides that Detroit experienced over these years, but we also know that autocorrelation is a likely problem. How should we check for this problem? As with heteroscedasticity, a plot of the residuals by the predicted values is a useful diagnostic tool.

As before, ask either Stata or SPSS to plot the studentized residuals by the predicted values from this regression model. The resulting plot, which can be found in Figure 11.7, indicates what appear to be heteroscedastic errors (note the drawn-out S-shape as we move from left to right), but is there evidence of autocorrelation? It is difficult to tell based on this plot, but try fitting a cubic or a lowess line. The cubic line indicates an autocorrelated structure. Moreover, the lowess line shows a huge bend in the lower left-hand portion of the plot and then some more curvature in the right-hand side of the plot. With additional experience, you will learn that this is also evidence of autocorrelated residuals.

As mentioned earlier, looking for patterns among the residuals and predicted values can be difficult, especially with large data sets. Fortunately, there are also statistical tests for autocorrelation. The most common of these is known as the *Durbin-Watson test.* It is calculated by considering the squared difference of the residuals adjacent in time relative to the sum of the squared residuals:

$$d = \frac{\sum_{2}^{n} (\hat{e}_{ti} - \hat{e}_{ti-1})^2}{\sum_{1}^{n} \hat{e}_{ti}^2}$$

The *t* subscript indicates time. This test is so common that most statistical programs, including Stata, compute it as part of their linear regression routines. A rule of thumb is that the closer the *d* value is to 2, the less likely that the model is affected by autocorrelation. The theoretical limits of d are 0 and 4, with values closer to 0 indicating positive autocorrelation (the most common type) and values closer to 4 indicating negative autocorrelation (a rare situation). To use the Durbin-Watson test, we first compute the d value and then compare it with the values from a special table of d values available in many regression textbooks and on the Internet. Some statistical programs provide values from these tables as part of their output. The values in the table include an upper limit and a lower limit based on the number of coefficients (explanatory variables + 1 or $\{k + 1\}$ and the sample size $[n]$). The following rules of thumb apply to the use of the d values from a linear regression model:

If $d_{model} < d_{lower}$, then positive autocorrelation is present.

If $d_{model} > d_{upper}$, then there is no positive autocorrelation.

If $d_{upper} < d_{model} < d_{lower}$, then the test is inconclusive.

Notice that this test does not assess negative autocorrelation, although, as mentioned earlier, if a d value from a model is close to the upper theoretical limit of 4, then there is negative autocorrelation that affects the regression model. Returning to the Detroit homicide example, we may ask Stata for the d statistic by first designating the time variable (`type tsset year`). Next, we can use the `estat dwatson` or `dwstat` post-command to tell Stata to calculate the Durbin-Watson statistic. If you are using SPSS, you can ask for the d statistic by checking the box labeled "Durbin-Watson" in the *Analyze-Regression-Linear-Statistics* menu. Both of these commands yield a Durbin-Watson of 0.66. This value is substantially lower than 2, but we should always check a table of Durbin-Watson values to determine whether it falls outside the boundaries for $\{k + 1\} = 2$ and $n = 13$. A table on the Internet indicates the following boundaries for a 5 percent test: {1.01, 1.34}. Hence, the results of the police–homicide linear regression analysis shows evidence of autocorrelation among the residuals.

Solutions for Autocorrelation

As with heteroscedasticity, autocorrelated errors are usually a symptom of incorrect model specification. Thus, a transformation of one or more of the variables may eliminate the problem. Unfortunately, it is difficult to find a suitable transformation in the presence of autocorrelation. Because we are concerned with correct specification, perhaps an important variable has been left out of the model. The most obvious possibility is a variable that gauges time. What happens in the Detroit example when we add *year* to the model? The d value increases to 1.23. The boundaries for a 5 percent test with $\{k+1\} = 3$ and $n = 13$ are {0.86, 1.56}. Because 1.23 falls within these boundaries, there is inconclusive evidence about autocorrelation. Nonetheless, looking at a scatter plot of the studentized residuals by the standardized predicted values indicates that there is still a problem with autocorrelation.

The best approach to solving the autocorrelation problem is to use a statistical technique designed for data that are collected over time. One such method is called time-series analysis, which is a substantial methodological approach that is beyond the scope of this presentation (a good resource for a fuller discussion is Ostrom, 1990). Both SPSS and Stata have the capacity to estimate time-series models, but it requires more background information than we have at present to use.

There are two popular regression-based approaches to diminishing the effects of autocorrelation: Prais-Winsten regression and Cochran-Orcutt regression. Both of these models use the lagged period of the data (for example, $t-1$) to come up with better estimates of the coefficients. In Table 11.3, we provide an example of a Prais-Winsten regression model (the second model shown) and compare it with the original OLS regression model (the first model shown). In Stata, this can be run using the command `prais`, whereas in SPSS you can use the SPSS Trends module.

The Prais-Winsten model has eliminated the autocorrelation problem. Notice that the D-W statistic is now well above the upper bound of 1.34, thus indicating that positive autocorrelation is no longer a problem. Moreover, we continue to find a significant association between the number of police and homicides in Detroit. Yet the slope is smaller, and the standard error is larger in the new model. A question remains, however: Why is the association positive?

There are also some useful techniques for other types of data collected over time, in particular longitudinal data. A set of techniques that is growing in popularity is known as generalized estimating equations (GEEs). One of the drawbacks of the Cochran-Orcutt and Prais-Winsten regression approaches is that they are limited to models with a *first-order autoregressive process*, or AR(1). This means that the immediate previous value has a much stronger direct influence than other preceding values on the current value (for example, 1970 → 1971). Yet it is often the case that other preceding values (for example, 1968, 1969) also have a strong effect on the current value. GEEs offer substantial flexibility to model various types of correlation structures among the residuals, not just AR(1). Stata has the capability to conduct GEE analyses.

Let's look at a couple of examples of GEEs. The *delinq* data set includes information collected over a four-year period from a sample of adolescents. It includes information on family relationships, self-esteem, stressful life events, and several other issues. Before considering a regression model, look at how the data are set up: Each adolescent provides up to four years of observations (if the ID number is the same, then it is the same adolescent). We can consider that there are two sample sizes: the number of people represented ($n = 651$) and the number of total observations, or $n \times t$ ($651 \times 4 = 2{,}904$). Moreover, a key assumption of the linear regression model is violated in this type of

Table 11.4: Comparison of Linear Regression and Generalized Estimating Equation [AR(1)] Models

	Linear Regression		Generalized Estimating Equation	
Variable	***b***	***SE***	***b***	***SE***
Cohesion	0.158	0.007	0.127	0.008
Stress	−0.376	0.047	−0.268	0.047
Constant	25.240	0.422	26.604	0.444
N	2,604		2,604	
R^2	.204		—	

Note: AR(1) = first-order autoregressive process.

data set: The observations are not independent because the same people contribute more than one observation.

Let's set up a linear regression model. We will predict self-esteem (*esteem*) among these adolescents using family cohesion (*cohes*; a measure of family closeness) and stressful life events (*stress*). A linear regression model provides the results shown in Table 11.4.

Now let's see what a GEE provides. Unfortunately, GEE models are not available in standard SPSS packages but require extensions, or you can use syntax files available on the Internet to run GEE models in SPSS. However, you can use SPSS's Advanced module to run some of these models using the *Linear Mixed Models* program. Because these modules are not standard in SPSS, we are going to use Stata to estimate these models. We first have to tell Stata that *year* is the time dimension and that the people in the data set are identified by the *id* variable. This involves using Stata's `xtset` command: `xtset id year`. Then the GEE model may be invoked: `xtgee esteem cohes stress, corr(ar1)`.

Note that the initial model is set up to assume that the errors follow an AR(1) process. Therefore, it is similar to a Prais-Winsten regression model. Note that the coefficients are closer to zero in the GEE model than in the OLS model. Hence, it is likely that the OLS model leads to biased coefficients when the observations are not independent. Autocorrelation does not affect the standard errors much in this model, though.

The AR(1) model may not be the best choice. Is it reasonable to assume that the value in previous years affects the current year's value but other years don't have much of an effect? Perhaps not; other years may also be influential. In a GEE model, we may estimate what is known as an *unstructured* pattern among the residuals. This allows the model to estimate the correlation structure rather than assuming that the residuals are uncorrelated (OLS) or follow an AR(1) pattern. The GEE model with an unstructured pattern, `xtgee esteem cohes stress, corr(unstructured)`, provides the results shown in Table 11.5.

Table 11.5: Comparison of Linear Regression and Generalized Estimating Equation (GEE) Models, with AR(1) and Unstructured Patterns

	Linear Regression		GEE AR(1)		GEE Unstructured	
Variable	***b***	***SE***	***b***	***SE***	***b***	***SE***
Cohesion	0.158	0.007***	0.127	0.008***	0.125	0.008***
Stress	−0.376	0.047***	−0.268	0.047***	−0.276	0.046***
Constant	25.240	0.422	26.604	0.444	26.697	0.441
N	2,604		2,604		2,604	
R^2	0.204		—		—	

Note: AR(1) = first-order autoregressive process.
***$p < .001$.

The first thing to notice is that there are few differences between the AR(1) model and the unstructured model. Therefore, the assumption of an AR(1) pattern among the residuals is reasonable. The `xtcorr` postcommand may be used to determine the correlation patterns from the models. However, keep in mind that if we do not take into consideration the longitudinal nature of the data, we overestimate the association between family cohesion and self-esteem and between stressful life events and self-esteem.

GEE models provide a good alternative to OLS linear regression models. When analyzing longitudinal data—which offer some important advantages over cross-sectional data—GEE models are the preferred approach. There are other models that are also popular, such as fixed effects models and random effects models, but these are actually specific cases of GEE models (a good discussion of these issues can be found in Hardin and Hilbe, 2003; and Singer & Willett, 2003).

A Brief Overview of Spatial Autocorrelation

Before concluding our discussion of autocorrelation, it is important to recall that another type is known as *spatial autocorrelation*. This problem occurs when the units in the data are spatial areas, such as counties, states, or nations. It should be evident that errors in prediction are likely to be more similar in adjacent or nearby units than in more distant units. Therefore, whenever spatial data are analyzed, it is a good idea to assess the likelihood of spatial autocorrelation because, as with serial correlation, the standard errors are affected. Moreover, as with longitudinal data, the assumption of independence is violated because adjacent areas usually share characteristics much more than distant areas (a more in-depth discussion of this issue is beyond the scope of this presentation, but Rogerson, 2006, is an excellent resource for this issue).

The Durbin-Watson statistic (d) is not useful for assessing spatial autocorrelation. However, a standard test is known as *Moran's I* (named after its originator, the Australian statistician Patrick A. P. Moran). Assuming we transform the variable into z scores (this simplifies the formula), Moran's I is calculated by the following equation:

$$I = \frac{n\sum_{i}^{n}\sum_{j}^{n} w_{ij} z_i z_j}{(n-1)\sum_{i}^{n}\sum_{j}^{n} w_{ij}}$$

In this equation, there are n spatial units, and w_{ij} is a measure of how close together the specific units, indexed by i and j, are to one another. If two spatial units that are close together exhibit similar scores on the variable (z), there is a positive contribution to Moran's I. Hence, if the nearest units tend to have scores that correlate more than units that are far apart, Moran's I is larger. In fact, its theoretical values are -1 and 1, much like Pearson's correlation coefficient, with higher values indicating positive spatial autocorrelation. Negative spatial autocorrelation also may occur, with Moran's I closer to -1, although it is rare. Notice that this measure is concerned conceptually with lack of independence across spatial units; hence, we may recognize autocorrelation as a problem of the dependence of observations.

An important issue in calculating Moran's I is to figure out the w_{ij}s because the I value is highly dependent on them. The simplest approach is to create a dummy variable with a value of one indicating that the spatial units are adjacent (or share a border) and a zero indicating that they are not. This is known as *binary connectivity*. Another way is to compute distance measures, although we then have to decide distances between which particular points. A commonly used distance measure is from the center of one region to the center of another (for example, the city center of Los Angeles to the city center of San Diego). If we use distance, w_{ij} in Moran's I is the inverse of the distance so that units closer together receive a larger weight.

As an example of how to compute Moran's I, consider the map in Figure 11.8. It consists of data from five counties in the state of Hypothetical. The numbers represent a measure of the number of crimes committed in each county in the last year adjusted for their population sizes. We wish to determine the degree of spatial autocorrelation before going any further in our analysis.

To compute Moran's I, we first need to decide on a system of weights for the distances. To ease the computations, let's use the binary connectivity approach where a one is assigned to adjacent counties (for example, A and C) and a zero is assigned to

Figure 11.8: Example of Spatial Autocorrelation

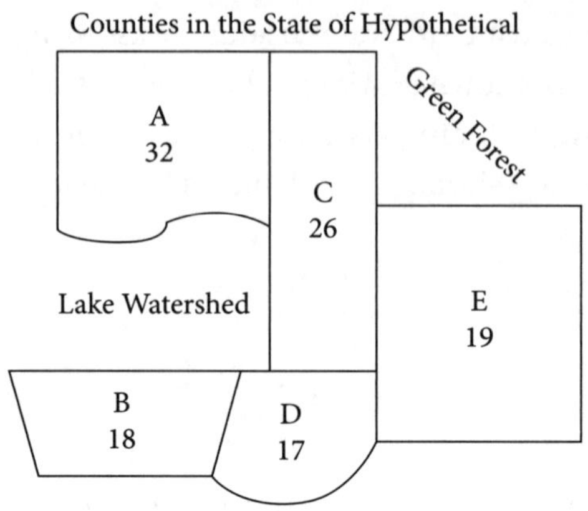

nonadjacent counties (for example, A and B). A simple way to see these weights is with a matrix where the entries are the w_{ij}s:

$$W = \begin{bmatrix} 0 & 0 & 1 & 0 & 0 \\ 0 & 0 & 0 & 1 & 0 \\ 1 & 0 & 0 & 1 & 1 \\ 0 & 1 & 1 & 0 & 1 \\ 0 & 0 & 1 & 1 & 0 \end{bmatrix}$$

This matrix is symmetric, with the same pattern above the diagonal as below the diagonal. For example, there is a 1 listed in row 1, column 3; and a 1 listed in row 3, column 1. This is an indicator that counties A and C share a border. The overall mean of the crime variable is 22.4, with a standard deviation of 6.43. To compute Moran's *I*, we may convert specific crime values into *z* scores to save us some steps. The *z* scores (calculations omitted), listed in order from counties A to E, are {1.494, –0.685, 0.56, –0.84, and –0.529}. We then add the products of each pair of *z* scores and multiply this sum by the sample size (5) to obtain the numerator (note that the nonadjacent pairs, because they have zero weights, are omitted from the equation):

$$\begin{aligned} &5 \times [AC + BD + CA + CD + CE + DB + DC + DE + EC + ED] \\ &\quad = 5 \times [(1.494 \times 0.56) + (-0.685 \times -0.84) + (0.56 \times 1.494) + (0.56 \times -0.84) \\ &\qquad + (0.56 \times -0.529) + (-0.84 \times -0.685) + (-0.84 \times 0.56) \\ &\qquad + (-0.84 \times -0.529) + (-0.529 \times 0.56) + (-0.529 \times -0.84)] = 10.9 \end{aligned}$$

The sum of the weights is 10 (count the ones in the **W** matrix), so the denominator in the equation is $\{4 \times 10\} = 40$. The Moran's *I* value is, therefore, $10.9/40 = 0.273$. This suggests that there is a modest amount of spatial autocorrelation among these crime statistics.

The solutions for spatial autocorrelation are similar in logic to the solutions for serial correlation. First, we may add a variable or set of variables to the model that explains the autocorrelation. It is often difficult to find these types of variables, though. Second, we may use a regression model—known as *geographically weighted regression*—that weights the analysis by a distance measure. Large weights apply to units that are close together; small weights apply to those that are farther apart. The regression coefficients are estimated in an iterative fashion after finding the optimal weight. Third, there are spatial regression models designed specifically to address spatial data and autocorrelation. The research area of *Geographical Information Systems* includes a host of spatial regression approaches. Rogerson (2006) provides a relatively painless overview of spatial statistics and regression models. Stata has several user-written programs available for spatial analysis (type `findit spatial models`). SPSS is less well suited to conduct such analyses.

Summary

We now know about the risks of heteroscedasticity and autocorrelation. Both of these problems are often symptoms of specification error in linear regression models, either because we use the wrong functional form or because we have left important variables out of the model. The main consequence of heteroscedasticity and autocorrelation is biased standard errors. Because we normally wish to obtain results that allow inferences to the population, obtaining unbiased standard errors is important. However, because specification error often accompanies heteroscedasticity or autocorrelation, the slopes from linear regression models may also be incorrect. Therefore, it is always essential that you think about the nature of your data and variables to evaluate the likelihood of specification error, lack of independent observations, heteroscedasticity, and autocorrelation.

There are several diagnostic tools and many solutions available for heteroscedasticity and autocorrelation. Here is a summary of several of these that are discussed in this chapter.

1. It is always a good idea to plot the studentized residuals by the standardized predicted values after estimating a linear regression model. A fan-shaped pattern to the residuals, one that either spreads out or narrows down at higher values of the predicted values, is indicative of heteroscedasticity. A cubic or similar snaking pattern to the residuals is indicative of serial correlation.
2. There are also several statistical tests for heteroscedasticity and autocorrelation. White's test and Breusch-Pagan test offer simple ways to use a regression model to test for heteroscedasticity. The Durbin-Watson d statistic is the most common numeric test for serial correlation.

3. Solutions to heteroscedasticity and serial correlation are legion. If you know the variable that is inducing heteroscedasticity or serial correlation, then you should consider a WLS regression analysis to adjust the model. If you don't know the source, or if you are genuinely worried about heteroscedasticity or serial correlation in your linear regression models, then use a model specifically designed to correct these problems. The Newey-West or Huber-White-Sandwich corrections for standard errors under heteroscedasticity work well. Prais-Winsten regression provides better estimates than OLS regression in the presence of serially correlated residuals.
4. For longitudinal data, a frequent approach is to use one of the GEE models. They provide a flexible way to model longitudinal associations among variables.
5. Data collected over spatial units present the same conceptual problems as data collected over time, although the techniques for diagnosing spatial autocorrelation and for adjusting the models are different. Moran's *I* is a widely used diagnostic test for spatial autocorrelation, and there are many regression routines designed specifically for spatial data.

Key Terms

The following is a list of key terms and concepts in chapter 11. You can find definitions to these words throughout the chapter.

- Autocorrelation
- Binary connectivity
- Breuch-Pagan test
- Durbin-Watson test
- First-order autoregressive process
- Geographical Information Systems
- Geographically weighted regression
- Heteroscedasticity
- Homoscedasticity
- Huber-White-Sandwich estimator
- Longitudinal data
- Repeated cross-sectional data
- Serial correlation
- Spatial autocorrelation
- Time-series data
- White's correction
- White's test

Analysis Exercises

1. For this exercise, we will use the *States* data set. Estimate a regression with *csat* as the dependent variable and the following as independent variables: *percent,*

expense, income, high, college, reg_1, reg_2, reg_3, reg_4, income2, percent2. Then answer the following:

a. Provide a graph that would indicate whether there is potential heteroskedasticity in the model. What do you conclude from this graph?
b. If the graph suggests that you should run a statistical test for heteroskedasticity, do so. If you run the heteroskedasticity test, what do you conclude from it? If you do not, why didn't you run the test?

2. For this exercise, use the *concord2* data set (these data come from the ATS Center at UCLA and were originally found in Hamilton, 1992). The time variable is called *time*. Estimate a regression with *H2Ouse* as the dependent variable, as predicted by *temp rain* and *educ*. Calculate the Durbin-Watson statistic for this regression. Does it show autocorrelation? Run a Prais-Winsten model and test for autocorrelation again.

Solutions

1. a. The graph is indeterminate.
   b. Both White's test and the Breusch-Pagan test suggest that there is no heteroscedasticity in the model.
2. The original regression shows positive autocorrelation because the model Durbin-Watson test of 0.535 was below the lower bound of the test. Running the Prais-Winsten changes the Durbin-Watson to 1.809, above the upper limit of the Durbin-Watson, indicating no positive autocorrelation.

CHAPTER 12 INFLUENTIAL OBSERVATIONS: LEVERAGE POINTS AND OUTLIERS

Chapter Objectives

In this chapter, we discuss how certain data points, outliers and leverage points, can affect regression models.

In chapter 12, we will cover

- what outliers and leverage points are and how to detect them
- various tests used to examine outliers and leverage points
- ways to deal with such data points.

We have seen many examples of scatter plots in previous chapters. These scatter plots show a variety of patterns for the relationships among variables. We have now seen linear patterns, nonlinear patterns, heteroscedastic patterns, and autocorrelated patterns. Yet nonlinear patterns, in particular, force us to question the use of a linear model to assess the association between two variables. It is not uncommon for two variables to be associated in a curvilinear or nonlinear fashion (see chapter 10). However, this does not exhaust the ways that data points diverge from a linear pattern. It is perhaps just as common to find small portions of the variables, perhaps even a single point, that diverge from the straight line. Even these small departures from linearity can have large consequences for the linear regression model.

In this chapter, we will discuss *influential observations*, or those data points that affect linear regression equations and results to an inordinate degree. The degree of influence is not always clear but can be quite substantial. There are two types of influential observations that affect a regression model: *outliers* and *leverage points*. Outliers are observations that pull the linear regression line in one direction or another. These are typically relatively extreme values of the y variable. In other words, they (there can be one or more) are far away from the other y values. For instance, suppose we have an outcome variable that measures the depression scores (on a scale of 0–40) from five high school seniors. We wish to predict these scores based on gender, family structure, and parental support (of course, we would want more than five observations to estimate such a model, but we will ignore this for now). Their depression scores are {5, 7, 11, 15, and 36}. The last score constitutes an outlier; it is extreme on the *y* variable.

Leverage points are observations that are discrepant or distant from the other values of the *x* variable. They may or may not also be outliers. For example, if we wish to use depression scores to predict grade point average (GPA), then the score of 36 would constitute a leverage point. The observation might also be an outlier if, say, the student who scored 36 had an unusually high GPA.

High leverage points that are not outliers mainly affect the standard error of the slopes. Recall that the formula for the standard error of a slope in a multiple linear regression model (OLS) is

$$\text{se}(\hat{\beta}_{\text{i}}) = \sqrt{\frac{\Sigma(y_i - \hat{y}_i)^2}{\Sigma(x_i - \overline{x})^2\left(1 - \text{R}_i^2\right)(n-k-1)}}$$

The key component of this equation that is affected by leverage points involves the x_i values in the denominator. Suppose that, all else being equal, we place a leverage point in the equation. What effect will it have? Well, it will result in a relatively large squared value in the denominator. Hence, the denominator will increase and the standard error will decrease. Thus, leverage points tend to reduce standard errors of slopes in linear regression models.

Figure 12.1 show an outlier, a leverage point, and a combination of the two. These graphs also include estimated linear regression slopes. Note that the outlier pulls what should be a relatively steep, positive regression line upward near the low end of the *x* distribution. Therefore, the slope will be smaller than if the outlier were not part of the set of observations. The leverage point falls on the regression line, so the slope is not influenced, but, as mentioned earlier, the standard error of the slope is affected. The bottom graph of Figure 12.1 shows a common situation: The observation is extreme on both the *x* and *y* variables. It has a relatively strong influence on the slope and the standard error. In brief, it has a large effect on the regression model.

Influential observations result from a number of sources. Perhaps the most common is coding error. When entering numeric data, it is easy to hit the wrong entry key or to forget a numeral or a decimal place. Therefore, it is always a good idea to explore the data to determine whether there are any coding errors. Often, though, influential observations occur as a routine part of data collection exercises: Some people do earn a lot more money than others; some adolescents do drink alcohol or smoke marijuana much more often than other youths. If the nature of a variable often leads to extreme values, then a common solution is to simply pull in these values by taking the square root or natural logarithm of the variable (see chapter 10). Before estimating a model, it is a good idea to use exploratory data analysis techniques, such as *q-q* plots, box plots,

Figure 12.1: Example of Outliers and Leverage Points

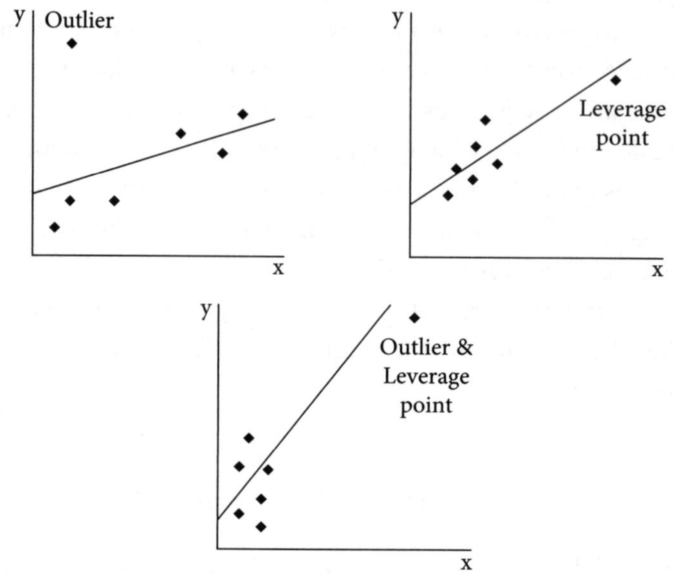

or stem-and-leaf plots, to visualize the distributions and check for leverage points or outliers.

Detecting Influential Observations

In a multiple linear regression context, there are several diagnostic techniques that are useful for unearthing influential observations. Here, we will give a brief overview of some of the more common techniques.

Detecting leverage points requires us to consider the spatial patterning among the sets of x values (assuming that we have multiple explanatory variables). If one observation is considerably discrepant from the others, it should show up as far from the others in a measure of spatial distribution. Recall that we may represent multiple linear regression models by using matrix notation (see chapter 3). Suppose we depict the x variables as a matrix with the entries representing particular observations:

$$\mathrm{X} = \begin{bmatrix} x_{11} & x_{12} & \cdots & x_{1p} \\ x_{21} & x_{22} & \cdots & x_{2p} \\ \vdots & \vdots & \cdots & \vdots \\ x_{n1} & x_{n2} & \cdots & x_{np} \end{bmatrix}$$

The *hat matrix* based on the **X** matrix is defined as

$$\mathrm{H} = \mathrm{X}(\mathrm{X}'\mathrm{X})^{-1}\mathrm{X}' = (h_{ij})$$

Recall that the matrix **X′** is the transpose of **X**, whereas $\mathbf{X}^{-1}$ is the inverse of **X** (see chapter 3). This is known as the hat matrix because $\hat{\mathrm{Y}} = \mathrm{HY}$, or the predicted values of *y* may be computed using the vector of *y* values premultiplied by **H**. The diagonals of the hat matrix (h_{ii}) lie between zero and one. We may compute the mean of these values to come up with an overall mean for the set of *x* values used in the multiple linear regression model. The points off the diagonals represent distance measures from this joint mean.

Many statistical software packages, including both SPSS and Stata, provide *leverage values* (sometimes called *hat values*) based on the hat matrix. Larger values are more likely to influence the results of the regression model. These values may range from zero to $\{n - 1\}/n$, which gets closer and closer to one as the sample size increases. A standard rule of thumb states that any leverage value that is equal to or exceeds $2(k + 1)/n$ should be scrutinized as an influential observation. For example, in a model with three explanatory variables and a sample size of 50, leverage points of $2(3 + 1)/50 = 0.16$ or more should be evaluated further.

The most common method for detecting outliers is through the use of *deleted studentized residuals* (also known as *jackknife residuals* or simply as *studentized residuals* by some researchers). These are very similar to the studentized residuals we used in earlier chapters (in Stata they are the same). In particular, they are estimated using the root mean square error ($\mathrm{S_E}$; see chapter 4) from the linear regression model after deleting the specific observation. The formula for deleted studentized residuals is

$$t_i = \frac{e_i}{\mathrm{S}_{\mathrm{E}(-1)}\sqrt{1-h_i}}$$

The e_i value is the unstandardized residual, and the h_i value is from the hat matrix described earlier. This is a standardized measure of the residuals, with a mean of zero, so we would expect, if the residuals are normally distributed, that about 5 percent of them would fall two units away from the mean. To detect outliers, look for deleted studentized residuals that are substantially greater than 2 or substantially less than −2.

As suggested earlier, it is not uncommon to find observations that are both leverage points and outliers. Hence, statisticians have developed a number of summary measures of influential observations. The two most prominent of these measures are Cook's *D* or distance (named for its developer, the statistician R. D. Cook) and DFFITS. Both of these measures combine information from leverage values (h_i) and deleted studentized residuals. Cook's *D* is computed as:

$$D_i = \frac{t_i^2}{k+1} \times \frac{h_i}{1-h_j}$$

Larger values of Cook's *D* indicate more influence on the regression results. There are various rules of thumb for Cook's *D*. One is to look for values greater than or equal to one. A more common—and less stringent—rule is to consider Cook's *D* values greater than $4/(n - k - 1)$ as indicative of influential observations. For example, in a model with three explanatory variables and a sample size of 75, any Cook's *D* value greater than $4/(75 - 3 - 1) = 0.056$ should be scrutinized.

The other general diagnostic measure of influential observations, DFFITS, is quite similar to Cook's *D*. It is computed as:

$$\text{DFFITS}_i = t_i \sqrt{\frac{h_i}{1-h_i}}$$

A general rule of thumb is to consider any DFFITS with an absolute value greater than or equal to two as indicative of an influential observation. (Note that DFFITS, like deleted studentized residuals, can be positive or negative.) However, a rule of thumb that considers the sample size is $2\sqrt{(k+1)/n}$; hence, if we have a model with three explanatory variables and a sample size of 75, any DFFITS values greater than $2\sqrt{(3+1)/75} = |0.46|$ should be evaluated as an influential observation.

Once these measures are estimated, it is a simple matter to compute the cutoff points for their respective rules of thumb and then identify influential observations. Another useful technique is to construct a scatter plot with the leverage values on the *x*-axis and the deleted studentized residuals on the *y*-axis. Then you may determine whether one or more observations are both outliers and high leverage points. A scatter plot of the Cook's *D* values (or the absolute value of the DFFITS) by the leverage values or deleted studentized residuals may also be useful for determining the type of influential observations that are affecting the model.

An Example of Diagnostic Tests for Influential Observations

Let's return to an example we used in earlier chapters: a multiple linear regression model with poor mental health days (*mntlhlth*) as the outcome variable. We will use the following explanatory variables: gender (*gender*), education (*educ*), and respondent income (*rincom06*). Estimate the model and predict the following: studentized residuals (`predict rstudent, rstudent`), Cook's *D*s (`predict cook, cook`), leverage

Table 12.1: Ordinary Least Squares Regression of Mental Health

Variable	*b*	*SE*
Gender	–0.921	0.478
Education	–0.199	0.084*
rincom06	–0.053	0.043
Intercept	7.837	1.151
N	963	
R^2	.016	

*$p < .05$.

Table 12.2: Descriptives of Diagnostic Measures

Variable	Observations	*M*	*SD*	Minimum	Maximum
Studentized	963	0.001	1.004	–0.862	3.888
Leverage	963	0.004	0.002	0.002	0.029
Cook's D	963	0.001	0.004	0.000	0.056
DFFIT	963	0.000	0.066	–0.148	0.478

values (`predict leverage, lev`), and DFFITS (`predict dfits, dfits`). The variable names for these predicted values are arbitrary. In SPSS, after you set up the model you can open the Save window and check the boxes next to Studentized deleted, Cook's, Leverage values, and Standardized DfFit. The results of the model are in Table 12.1. Summary statistics for the diagnostic measures we predicted are found in Table 12.2.

The results provide the minimum and maximum values, but we are mainly interested in the maximum values of the leverage and Cook's D measures. First, however, it is useful to compute the cutoff values for these measures. Given our model, the leverage value rule-of-thumb cutoff is 2(3 + 1)/963 = 0.008; the Cook's *D* rule-of-thumb cutoff is 4/(50 – 3 – 1) = 0.004; and the DFFITS rule-of-thumb cutoff is $2\sqrt{(3+1)/968} = |0.129|$.

According to Table 12.2, the maximum studentized residual value is 3.888, the maximum leverage value is 0.029, the maximum Cook's *D* value is 0.056, and the maximum DFFITs value is 0.478 (in absolute value terms). All tests show some cause for concern. The studentized residuals show at least one observation much higher than two units from the mean of zero. Further, there is at least one observation that exceeds the preferred maximum leverage, Cook's *D*, and DFFITS values.

A convenient way in Stata to consider these values in more detail is by using additional exploratory analysis techniques. For example, by using the `list` command to list all Cook's *D* and leverage values greater than the cutoff values (`list cook if cook > 0.004` and `list leverage if leverage > 0.008`), we may identify which cases comprise the maximum values for these diagnostic variables. In SPSS, we can look at the *Extreme Values* output for further information. For example, the largest Cook's *D* value (0.056) is individual 107; the next largest is case number 1441 (Cook's $D = 0.046$). According to the cutoff value, both of these cases are influential observations, along with several others. The largest leverage value (0.029) is case 454. According to the rule of thumb for the leverage values, there are several leverage points that exceed the rule of thumb. Box plots and stem-and-leaf plots are also useful for exploring data. There are six individuals that appear as an influential observation in all three diagnostics, so we should look at these observations in greater detail.

Solutions for Influential Observations

We have already discussed one of the best solutions for influential observations: Make sure the variables are coded correctly. Often, influential observations are due to coding errors or nonnormally distributed variables. Carefully exploring the variables before estimating a regression model through the various techniques discussed in earlier chapters will go a long way toward alleviating problems with influential observations.

Though rarely recommended, another solution is to delete the influential observation. For example, if we cannot figure out why these six observations are affecting our regression model, then we might simply remove them from consideration. For example, Table 12.3 provides the results of the regression model without these observations. The results change quite dramatically. In particular, the unstandardized coefficient for gender is more negative and became statistically significant. Nonetheless, this shift in the coefficient should not be used as an excuse to remove the influential observation, but it does reveal a clue that we will consider later.

As an alternative, we could simply ignore the influential observations. In a large sample, such as ours, this may be a reasonable alternative (although the solutions discussed later are preferred). For smaller samples, though, influential observations can have a large impact on the results.

The best solution is to try to understand influential observations. For example, why do these six observations have such an effect on the results? Let's explore this some more by considering the explanatory variable—gender—that seems most affected by these observations. All six respondents are male, and all six have outlier values on the mental health variable. In fact, three of the respondents reported 30 poor mental health days in the past month—the highest possible value. Another respondent reported 25 poor days, while the other two reported 15 and zero days. These influential observations may have a large effect as a result of these values. However, because gender is a

Table 12.3: Ordinary Least Squares Regression of Mental Health without Influential Observations

Variable	*b*	*SE*
Gender	−1.225	0.470**
Education	−0.203	0.085*
rincom06	−0.025	0.043
Intercept	7.506	1.145
N	957	
R^2	.016	

*$p < .05$. **$p < .01$.

dichotomous variable, we do not have many options for correcting the variable to eliminate our influential observation problem. One possibility is transforming the variable for mental health. If we look at the summary statistics for this variable, we can tell it is highly skewed (median = 0, mean = 3.825, skew = 2.427). After doing so and reestimating the measures for leverage and influential observations, we notice that we still have some large leverage, Cook's *D*, and DFFITS values. Thus, we did not show much improvement after specifying the model in a different way.

Sometimes people solve the problem of influential observations by using regression techniques that are less sensitive to extreme values. For instance, you may recall that the median is known as a robust measure of central tendency because it is less sensitive than the mean to extreme values (see chapter 1). Because our multiple linear regression model using least squares is based on means, perhaps an alternative regression technique based on medians would be influenced less by extreme values. This reasoning is the basis for two very similar regression models: *median* regression and *least-median-squares* regression. Rather than minimizing the sum of squared errors, these techniques minimize the *sum of absolute residuals* or the *median squared residual.* Such models are more easily run in Stata than in SPSS. This can be done using the `qreg` command. However, you will need a regression model with only continuous variables, because dichotomous variables do not technically have medians.

Another alternative is to use the Huber-White-Sandwich estimators (see chapter 11) to adjust the standard errors for influential observations. Recall that this estimation technique is useful for heteroscedastic errors. Often, influential observations mimic heteroscedasticity, so adjusting the standard errors with this widely used approach yields more robust results. The results of an OLS model with Huber-White-Sandwich estimators of the standard errors are presented in Table 12.5.

A related approach is known as *robust regression* (there are actually several types, including median regression). One form of robust regression begins with an OLS

Table 12.4: Results for Exercise 2

	OLS		MLS	
Variable	***b***	***SE***	***b***	***SE***
Age	0.073	0.011***	0.071	0.014***
Education	0.759	0.056***	0.857	0.070***
Intercept	0.348	0.915	0.143	1.136

Note: OLS = ordinary least squares; MLS = median least squares.
***$p < .001$.

Table 12.5: Results for Exercise 3

	OLS		MLS		Robust	
Variable	***b***	***SE***	***b***	***SE***	***b***	***SE***
Age	0.073	0.011***	0.071	0.014***	0.085	0.012***
Education	0.759	0.056***	0.857	0.070***	0.814	0.058***
Intercept	0.348	0.915	0.143	1.136	−0.583	0.950

Note: OLS = ordinary least squares; MLS = median least squares.
***$p < .001$.

model and then calculates weights based on the absolute value of the residuals. It goes through this process repeatedly (*iteratively*) until the change in the weights drops below a certain point. In essence, it down-weights observations that have the most influence on the model. SPSS does not include a robust regression technique as part of its regular modeling routines. However, Stata includes a robust regression technique as part of its regular suite of modeling routines, using the `rreg` (robust regression) command.

The results of these various models indicate that the statistical significance of the gender variable depends on the method we are using. Moreover, even though the evidence is clear that there are influential observations, a number of robust techniques suggest that the effects of these influential observations would not force us to change our general conclusions. However, some of the robust methods lead to a significant gender variable, and others do not. Removing the influential observations leads to a significant gender effect as well. Because the sample size is so large here, it may be fine to leave the influential observations in the model, but each of the methods we discussed have pros and cons associated with them.

We have touched only the surface of the abundant information about influential observations and the many robust regression techniques that are available. Several books and articles provide good overviews of several robust regression techniques. The main point is that influential observations are a common part of regression models. Although they are often caused by coding errors or reflect highly skewed variables, they are also a routine part of data collection and variable construction. Exploring your data carefully before estimating the model and assessing influence measures after estimating the model should therefore be part of any regression exercise.

Key Terms

The following is a list of key terms and concepts in chapter 12. You can find definitions to these words throughout the chapter.

- Deleted studentized residuals
- Least-median-squares regression
- Leverage points
- Leverage values
- Outliers
- Robust regression

Analysis Exercises

1. Estimate a regression predicting personal income (*rincom06*) as a function of age (*age*) and education (*educ*) and answer the following questions:
   a. What are the critical values for leverage, Cook's *D*, and DFFITS in this regression?
   b. How many respondents have leverage, Cook's *D*, and DFFITS values above the critical value for all three measures?
2. Estimate a median regression model: How do the results differ from the standard OLS regression model?
3. Estimate a robust regression model: How do the results differ from the standard OLS and median regression model?

Solutions

1. a. For leverage, the value is .005; for Cook's *D*, it is .003; and for DFFITS, it is .10.
   b. Eight respondents have values above all three critical values.
2. See Table 12.4.
3. See Table 12.5.

CHAPTER 13 AN INTRODUCTION TO LOGISTIC REGRESSION

Chapter Objectives

In this chapter, we introduce the logistic regression model. This model is used when we have a binary (0–1) outcome, which means that linear regression cannot be used.

In chapter 13, we will

- introduce logistic regression and discuss why it is a better option than linear regression for binary outcomes
- highlight how logistic regressions are fit and estimated
- discuss how to interpret results from a logistic regression
- show how to run logistic regression in Stata and SPSS.

Although this book has emphasized outcome variables that are continuous and either normally distributed or transformable to normality, it is actually the rare research endeavor that uses only these types of variables. It is much more common to find outcome variables that are measured as categories, such as low, medium, and high; or yes and no. The topic of categorical outcome variables requires a full book to examine the many modeling possibilities. Therefore, in this chapter, we will briefly introduce and discuss perhaps the most popular of the categorical variable regression techniques: *logistic regression*. To be precise, this model is identified as *binary logistic regression* because it is concerned with binary outcome variables.

There are many examples of binary variables (also known as dichotomous variables) in the social, behavioral, and health sciences. We have seen some examples when discussing dummy variables: Gender, race/ethnicity, religious preferences, and many others may be conceptualized as one or more two-category variables. It is not uncommon to find binary variables as outcome variables. For example, some binary outcome variables that appear in the research literature include support for the death penalty, graduation from college, whether companies in a sample test for illegal drugs, whether people vote, whether Republican or Democratic presidential nominees win a

Figure 13.1: Example of Why Ordinary Least Squares Is Inappropriate for Dichotomous Outcomes

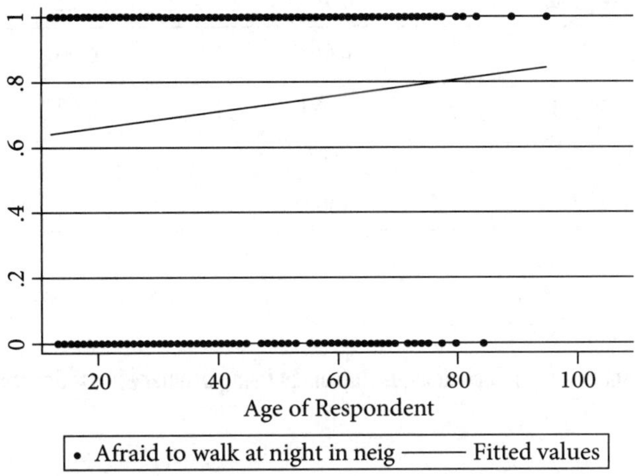

Note: neig = Neighborhood.

state, whether adolescents drink alcohol, and whether a medicine cures an illness. All of these variables lend themselves, some better than others, to yes-or-no responses.

It should be clear that this type of variable violates at least two of the key assumptions of multiple linear regression models. In chapter 3, we learned that one of these assumptions states: The error term is normally distributed with a mean of zero and constant variance. Another assumption declares: The mean value of y for each specific combination of the xs is a linear function of the xs. Now, imagine a binary outcome variable in a model; can it satisfy either of these assumptions? The quick answer is no. To gather evidence to this effect, consider one of the variables that appears in the GSS 2010 data set. The variable *fear* measures whether respondents are afraid to walk alone in their neighborhood at night. The values are no (coded 0) and yes (coded 1). Of course, it would be preferable to have a continuous measure of fear, but suppose this is all we have. Can we assess predictors of this variable in a regression context? For instance, does age predict neighborhood fear?

As an initial step, we can ask either Stata or SPSS to construct a scatter plot with neighborhood fear (*fear*) on the y-axis and age in years (*age*) on the x-axis. Place a linear fit line in the scatter plot. You should find a plot with a fit line that looks like Figure 13.1.

Simply looking at the fit line, we would conclude that there is a positive association between age and neighborhood fear. However, notice how peculiar the scatter plot looks. All the fear scores fall in line with the values of zero or one. This, in itself, is not peculiar because respondents can take on scores of only zero or one. However,

Table 13.1: Ordinary Least Squares Regression for Neighborhood Fear

Variable	*b*	*SE*
Age	0.003	0.001*
Intercept	0.591	0.053
N	777	
R^2	.008	

*$p < .05$.

Figure 13.2: Nonlinearity in Dichotomous Outcomes Using Ordinary Least Squares Regression

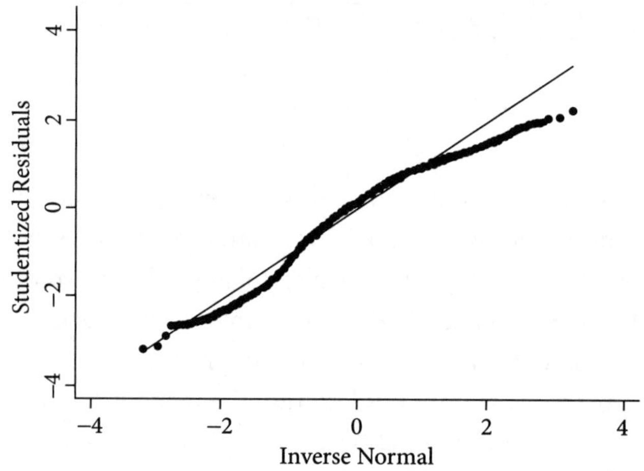

try to imagine lines running from the points to the fit line. Notice that there will be a systematic pattern to these residuals. Is this indicative of any problems we have discussed in previous chapters? The issue is that there is a systematic pattern that indicates heteroscedasticity.

As a next step, let's estimate a linear regression model with fear as the outcome variable and age as the explanatory variable. When setting up the model, ask for a normal probability plot of the residuals. The results (see Table 13.1) provide an unstandardized coefficient of −0.003. Hence, the interpretation is that each one-year increase in age is associated with a 0.003-unit increase in neighborhood fear. Does this interpretation make sense? Can neighborhood fear increase by 0.003 units? This illustrates a key problem with using a linear regression model to predict a binary outcome variable. Look at the normal probability plot of the residuals in Figure 13.2.

Is it even possible to claim that the residuals approach a normal distribution? This would constitute a leap in interpretation no sincere analyst would be willing to make. Another problem is that, even though these types of outcome variables can take values of only zero or one, it is possible to obtain predicted values from a linear regression equation that exceed one or are less than zero.

An Alternative to the Linear Model: Logistic Regression

One assumption that has appeared in the statistical literature is that, as long as the binary outcome variable is distributed somewhere close to 50 percent in one category and 50 percent in the other (some claim that anything between 30 percent and 70 percent is acceptable), then the linear regression model—which is labeled a *linear probability model* when it includes a binary outcome variable—yields acceptable results. However, there is no longer any reason to rely on this model as there are several widely available and easy-to-use alternatives. We will discuss the most popular alternative model used in the social, behavior, and health sciences: binary logistic regression. As background, consider that, when interested in a nonnormal outcome variable, one of the tools we used was a transformation. A variable with a long right tail, for example, is typically transformed using the natural logarithm or square root function. Although not as apparent from a mathematical point of view, we may also transform a binary outcome variable by using a specific function. Before seeing this function, it is important to recall some information from elementary statistics.

Remember when you were asked in some secondary school mathematics course to flip a coin and estimate elementary probabilities? You were probably told that, given a fair coin, the probability of a heads was 0.50 and the probability of a tails was also 0.50. That is, from a frequentist view of statistics, we expect that if we flip a coin numerous times, about half the flips will come up as heads and the other half will come up as tails. This is usually represented as $P(\text{H}) = 0.50$, with P shorthand notation for probability (see chapter 1). The development of logistic regression—as well as other techniques for binary outcome variables—was predicated on the notion that a binary variable could be represented as a probability. Returning to our earlier example, we might ask what the probability is of people reporting that they do not or do fear walking alone in their neighborhood. Using our shorthand method, this is depicted as $P(\text{no fear}) = 0.29$. In practical terms, this means that about 29 percent of the sample respondents report no fear of walking alone at night. A fundamental rule of probabilities is that they must add up to one. Hence, if there are only two possibilities, we know that the probability of fear $\{P(\text{fear})\}$ is $1 - 0.29 = 0.71$. The simplest way to determine these probabilities is to ask SPSS or Stata for frequencies for the *fear* variable.

Table 13.2: Original and Transformed Values

Initial Value	Transformed Value
−10.0	0.000045
−5.0	0.006693
−0.5	0.487503
0	0.5
0.5	0.622459
5	0.993307
10	0.999955

Once we shift our interest from estimating the mean of the outcome variable to estimating the probability that it takes on one of the two possible choices, we may use a regression model that is designed to predict probabilities rather than means. This is what a binary logistic regression model is designed to do: It estimates the probability that a binary variable takes on the value of one rather than zero (notice that, like dummy variables, we assume that the outcome variable is coded {0, 1}). In our example, we may use explanatory variables—which may be continuous or dummy variables—to estimate the probability that people are fearful.

Binary logistic regression accomplishes this feat by transforming the linear regression equation through the use of the following *logistic function*:

$$P(Y=1)=\frac{1}{1+\exp[-(\alpha+\beta_1 x_1+\beta_2 x_1+\cdots+\beta_k x_k)]}$$

The part of the denominator in parentheses is similar to the linear regression model that we've used in earlier chapters. However, note that it is transformed in a very specific way. The advantage of this function is that it guarantees that the predicted values range from zero to one, just as probabilities are supposed to do. For example, Table 13.2 shows several negative and positive predicted values (that is, from the linear regression equation in the denominator) and their values after running them through the function. If we placed more extreme values in the function, it would return numbers closer to zero or one, but they would never fall outside the boundary of [0, 1].

Logistic regression models do not use OLS or some similar estimation routine. Rather, they use a widely used estimation procedure known as *maximum likelihood* (ML). ML is a common statistical procedure, but its particulars require more statistical knowledge than we currently have at our disposal. Suffice it to say that ML estimates the most likely value of a statistic, whether the value of interest is a mean, a regression coefficient, or a standard error, given a particular set of data. More detailed information on ML estimation is available in most books on statistical inference.

Fortunately or unfortunately (we will let you decide), logistic regression models are not often used to come up with predicted probabilities. Although there is no good reason why this occurs, most researchers who use logistic regression use *odds* and *odds ratios* to understand the results of the model. Odds are used frequently in games of chance. For example, slot machines and roulette wheels are usually "handicapped" by being accompanied by the odds of winning any particular attempt. Horse races are also supplemented by odds: "The odds that Felix the Thoroughbred will win the Apple Downey Half-Miler are four to one." In general, odds are simply probabilities that have been transformed using the following equation:

$$\frac{P}{(1-P)}$$

Say that our thoroughbred has been in 10 half-mile races and won two of them. We may then say that the probability that she wins a half-miler is 0.20. This translates into an odds of winning of $0.20/(1-0.20)=0.25$. Another way of saying this is that the odds of winning are four to one, or that for each race she wins, she loses four of them.

An example closer to the interests of the research community is the following. Suppose we conduct a survey of adolescents and find that 25 percent report marijuana use and 75 percent report no marijuana use in the past year. Hence, the probability of marijuana use {*P*(marijuana use)} among adolescents in the sample is 0.25. What are the odds of marijuana use?

$$\frac{P(\text{marijuana use})}{(1-P\{\text{marijuana use}\})}=\frac{0.25}{1-0.25}=0.33 \text{ or } 1/3$$

This implies that for every three adolescents who do not use marijuana, we expect one adolescent to use marijuana. In other words, three times as many adolescents have not used marijuana as have used marijuana in the past year.

This is not very interesting, though, as we have restricted our attention so far to only one variable. Let's extend our attention to two variables. We will treat adolescent past-year marijuana use as the outcome variable. To keep it simple, we will use a dummy variable that measures gender (coded 0 = female; 1 = male). Now we have up to four outcomes, with males and females who have used or have not used marijuana. If interested in probabilities, we may compare the two: *P*(male marijuana use) and *P*(female marijuana use). However, in keeping with our interest in odds, we will compare the odds of marijuana use among male and female adolescents.

Let's assume that our survey data indicate that 40 percent of male adolescents and 20 percent of female adolescents report past-year marijuana use. Hence, the odds of marijuana use among males is 0.40/0.60 = 0.667 or 2/3, and the odds of marijuana use among females is 0.20/0.80 = 0.25 or 1/4. Keep in mind that these are not probabilities; odds are different. Another way of saying this is that for every two males who used marijuana, three did not; and for every one female who used marijuana, four did not.

An odds ratio is just what it sounds like: the ratio of two odds. Let's continue our adolescent marijuana use example to see the utility of this measure. Assume that we wish to compare marijuana use among males and females. What is the odds ratio for these two groups? First, we have to decide on which is the focal group and which is the comparison group. Let's use males as the focal group, as they are more likely to report use. The odds ratio (*OR*) of males to females is

$$OR_{\text{males vs. females}} = \frac{\text{Odds (males)}}{\text{Odds (females)}} = \frac{P(\text{males})}{1-P(\text{males})} \bigg/ \frac{P(\text{females})}{1-P(\text{females})} = \frac{0.667}{0.25} = 2.67$$

An odds ratio of 2.67 gives rise to the following interpretation: The odds of past-year marijuana use among males are 2.67 times the odds of past-year marijuana use among females. Some analysts use a shorthand phrase that males are 2.67 times *as likely* as females to use marijuana. However, this can mislead some observers into thinking that we are dealing with probabilities rather than odds. The ratio of the probabilities is 0.40/0.20 = 2.0, which is clearly not the same as the odds ratio.

Odds ratios are simply two odds that are compared to determine whether one group has higher or lower odds than another group on some binary outcome. A number greater than one implies a positive association between an explanatory and outcome variable (but keep in mind that the coding of the variables drives the interpretation) and a number between zero and one indicates a negative association. If there is no difference in the odds, then the odds ratio is one. For instance, if males and females are equally likely to report past-year marijuana use, then the odds ratio is 1.0.

A logistic regression model is one way to compute predicted odds and odds ratios for binary outcome variables. A useful equation that lends itself to these computations is

$$\text{logit}[P(y=1)] = \alpha + \beta_1 x_1 + \beta_2 x_2 + \cdots + \beta_k x_k,$$

$$\text{where the logit}[P(y=1)] = \log_e\left[\frac{P(y=1)}{1-P(y=1)}\right]$$

The quantity $\log_e[\frac{P(y=1)}{1-P(y=1)}]$ is also known as the *log-odds*. The term $\log_e$ indicates the natural logarithm. Recall the logistic function that was specified earlier. It is another version of this equation. Moreover, the term *log-odds* should bring to mind a possible

Table 13.3: Cross-tabulation of Gender and Fear

	Gender		
Fear	**Male**	**Female**	**Total**
No	147	75	222
Yes	244	311	555
Total	391	386	777

solution to computing the odds: Take the exponential of this quantity to transform it from log-odds to odds (or, as we shall see, an odds ratio).

This can be a confusing exercise, so let's return to a real example of a binary outcome variable. In the GSS 2010 data we used earlier, there is a variable *gender* that is coded as 0 = female and 1 = male. We will treat *gender* as the explanatory variable and *fear* as the outcome variable. The results of a cross-tabulation for these two variables is shown in Table 13.3.

What are some of the relevant probabilities and odds from this table? First, the probability of fear overall is 555/777 = 0.714. The overall odds of fear are 555/222 = 2.5 (notice that we do not need to compute probabilities to compute odds). These overall probabilities and odds may be interesting, but they are not very informative. Rather, we may also consider specific measures among males and females. The probabilities of fear among males and females are 244/391 = 0.624 and 311/386 = 0.806, respectively. The respective odds are 244/147 = 1.66 among males and 311/75 = 4.147 among females. We can easily see that females are more likely than males to report feeling fearful in their neighborhood.

The odds ratio provides a summary measure of the association between gender and fear. Because we have the two odds, computing the odds ratio is simple: $OR_{\text{females vs. males}} = 4.147/1.66 = 2.498$. An interpretation is that the odds of neighborhood fear among females is 2.498 times the odds of fear for males. Recall that an odds ratio greater than 1.0 indicates a positive association. Because females constitute the higher coded category, we claim a positive association between gender and fear (without, of course, making any qualitative judgments about gender and fear).

It appears that there is little advantage to going beyond this simple exercise to estimate the association between gender and fear. However, keep in mind that we might wish to (1) determine whether there is a statistically significant association between gender and fear (assuming that we have a good sample from a population) and (2) add additional explanatory variables to determine whether the association is spurious. This is where binary logistic regression comes in handy: It estimates standard errors and p values; and it allows additional explanatory variables in the model.

Table 13.4: Logistic Regression of Gender and Fear

Variable	*OR*	*SE*
Gender	2.498	0.414***
Intercept	1.660	0.173

***$p < .001$.

First, we will estimate the simple logistic model with one explanatory variable. In Stata, this model is requested with the `logit` command, while in SPSS it can be done by using the drop-downs: *Analyze-Regression-Binary Logistic.* The `or` option converts the coefficients into odds ratios (as an alternative, the `logistic` command automatically provides odds ratios) in Stata, while SPSS automatically offers the odds ratios. Table 13.4 shows our results.

The appearance of the results is similar to that obtained using a linear regression model, with a couple of important exceptions. The column labeled *OR* is of particular interest. The value in the Gender row should look familiar: It is the same value we obtained for the odds ratio from the cross-tabulation (but keep in mind that this works only because we coded female as 1 and male as 0). In fact, it is the odds ratio for the gender–fear association. However, assuming that we wish to infer something about the population from which the sample was drawn, we now may determine whether this odds ratio is significantly different from 1.0. The p value is used in the same manner as in the linear regression model, with, for instance, a value below .05 typically recognized as designating a statistically significant result. According to the p value of less than .001, we may claim that the odds ratio is significantly different from 1.0 at the $p < .001$ level and conclude (tentatively, in the absence of much more information) that females are more likely to report feeling fearful in their neighborhood. Using the information on significance tests in regression models (see chapter 2), how should we interpret a p value this small?

The interpretation of the odds ratio is the same as it was earlier: The odds of women feeling fearful in the neighborhood are expected to be 2.498 times the odds of men feeling fearful. Notice that we have added the phrase "expected to be" rather than something more determinate. This is to remind the reader that we are inferring from a sample to a population. We can never be certain, given a single sample, that the odds among females *are* 2.498 times higher—only that we expect or infer them to be this much higher given the model results.

Suppose we wish to figure out the odds of fear among males and females rather than just the summary odds ratio. After reestimating the model without odds ratios, we may use the log-odds ratios to estimate predicted values. For example, the predicted value for males is [.507 + {.915 × 0}] = .507, and the predicted value for females is {.507 + [.915 × 1]} = 1.422. Recall the equation presented earlier in the chapter:

$$\text{logit}[P(y=1)] = \alpha + \beta_1 x_1 + \beta_2 x_2 + \cdots + \beta_k x_k,$$

$$\text{where the logit}[P(y=1)] = \log_e\left[\frac{P(y=1)}{1-P(y=1)}\right]$$

Notice that the values we just predicted are therefore known as logit-values or, more commonly, the log-odds. Stata predicts these values in a logistic regression model when the `or` option is not included as it was in the previous model. To transform the log-odds into odds, we simply take the exponential (inverse of the natural logarithm) of these values. The predicted odds for males are thus *exp*(.507) = 1.66 for males and *exp*(1.422) = 4.145 for females. Compare these with the odds we computed using the cross-tabulation of gender and neighborhood fear.

There are some—perhaps many—researchers who are uncomfortable with odds and odds ratios and prefer probabilities. The coefficients from a logistic regression model may be transformed into probabilities by using the logistic function:

$$P(y=1) = \frac{1}{1+\exp[-(\alpha + \beta_1 x_1 + \beta_2 x_2 + \cdots + \beta_k x_k)]}$$

We simply substitute the part of the denominator in parentheses with the logistic regression coefficients. However, we must choose specific groups to represent. With the previous model, there are only two groups, females (coded 0) and males (coded 1), and we may predict their probabilities by including the coefficients in the equation:

$$P(\text{females}) = \frac{1}{1+\exp[-(-0.26+\{1.216\times 0\})]} = 0.435$$

$$P(\text{males}) = \frac{1}{1+\exp[-(-0.26+\{1.216\times 1\})]} = 0.722$$

Notice that these predicted probabilities are identical to the probabilities we computed with the cross-tabulation of gender by neighborhood fear. In Stata, you may also save the predicted probabilities and then assess their values for particular groups that were included in the model. However, Stata provides even better tools for estimating probabilities. The `adjust` postcommand, for example, may be used to predict probabilities for particular values of the explanatory variables. A user-written program, `prvalue`, is also useful (type `findit prvalue`). Here is an example of the `margins` command to request probabilities (the `adjust` postcommand may also be used):

```
margins, at(gender=0)
margins, at(gender=1)
```

Stata returns the probabilities of .624 and .806, which, as expected, are the same as computed previously.

In SPSS, you can save the predicted probabilities as well. The simplest way to do this in SPSS is to save them by opening the Save window in the Binary Logistic menu and checking the appropriate box. Then use the *Analyze-Compare Means-Mean* drop-down menu with the predicted probabilities (labeled PRE_1) in the Dependent List box and *gender* in the Independent List box. However, note that this works best with dummy variables or independent variables with a few categories. In both SPSS and Stata, there are also options for residuals and influence statistics that we will ignore in this presentation but are worth exploring and understanding if you wish to realize the capabilities and limitations of specific logistic regression models.

Extending the Logistic Regression Model

We have not reached the limit of the binary logistic regression model. Rather, as with the linear regression model, we may estimate models with multiple explanatory variables—whether dummy or continuous—and interaction terms. Let's throw a multiple logistic regression model together. We will continue to use *fear* as the outcome variable and include the following explanatory variables: *gender, age,* and *income* (*rincom06*). We will treat *age* and *income* as continuous variables. The results of this model are in Table 13.5.

The model indicates that, statistically adjusting for the effects of age and income, the odds of fear for females is expected to be 2.353 times the odds of fear among males. This is also known as an *adjusted odds ratio* because we have adjusted for the presumed effects of other variables in the model. Note, though, that adjusting for these effects has little influence on the association between gender and fear. Moreover, we may interpret the odds ratio for age or income. Let's focus on age because it is at least close to the normal threshold ($p < .05$) for statistical significance. Given that it is a continuous variable measured in year increments, the interpretation is that *statistically adjusting for the effects of gender and income, each one-year increase in age is associated with a 1.011 times difference in the odds of neighborhood fear.*

This type of interpretation is not very satisfactory to most analysts. It is difficult, without substantial experience, to interpret this as a positive association between age and fear. However, there is a nice property of regression models that transform outcome

Table 13.5: Logistic Regression of Neighborhood Fear

Variable	*OR*	*SE*
Gender	2.353	0.169***
Age	1.011	0.006
rincom06	1.022	0.014
Intercept	0.778	0.299

***$p < .001$.

variables using natural logarithms that is useful in this situation. That is, we may use the following *percentage change formula* to interpret coefficients from logistic regression models:

$$\{\exp(\beta) - 1\} \times 100$$

This formula uses the untransformed coefficient to determine the percent difference or change in the odds associated with a one-unit difference or change in the explanatory variable. Using the age coefficient and this formula, we find $\{[1.011] - 1 \times 100\} = 1.1\%$ (we do not need to exponentiate the coefficient because it is already an odds ratio). Thus, the interpretation of the coefficient is that *statistically adjusting for the effects of gender and income each one-year increase in age is associated with a 1.1 percent increase in the odds of feeling fearful.*

Multiple logistic regression models may also be used to estimate predicted odds or probabilities. This is complicated by the presence of continuous variables, however. As with multiple linear regression models (see, for example, chapter 6), we should choose particular values of the explanatory variables to compare. Some researchers prefer to use minimum and maximum values, but any plausible values, as long as they are represented by a sufficient number of cases in the data set, work reasonably well.

Finally, there are tests of fit for logistic regression models that are analogous to fit statistics in linear regression models. Nonetheless, these are not computed in the same way, nor should they be interpreted in the same way as F values, S_E values, or R^2 values. They may, however, be used to compare models. For example, Stata provides a log likelihood statistic, which may be used to calculate the *deviance* statistic ($-2 \times \log$ likelihood). The deviance statistic may be used to compare nested models (but not non-nested models). Values closer to zero tend to indicate better fitting models, although additional information is required before one may reach this conclusion. As an example, the original model with only gender as an explanatory variable has a deviance of 31.85, whereas the model with gender, age, and income has a deviance of 39.20. To compare these models, subtract one from the other. The result is a distributed chi-square with

Figure 13.3: Comparison of Two Models Using Stata's Test Postcommand

```
test age rincome
( 1) [fear]age = 0
( 2) [fear]rincome06 = 0

     chi2( 2) =   7.31
     Prob > chi2 =   0.0259
```

degrees of freedom equal to the difference in the number of explanatory variables across the two models. Therefore, one takes the difference between the two deviances—in this case, 39.20 – 31.85 = 7.35—and compares it with a chi-square value with two degrees of freedom (because we added two explanatory variables in the second model). The *p* value associated with this chi-square value is 0.062—in Stata, type display 1 - chi2(3,7.35)—which suggests that the difference between the two models is not statistically significant. In other words, we do not improve the model's predictive capabilities by including age and income. In the spirit of parsimony, we should conclude that the simpler model provides a better fit to the data.

A simple way in Stata to obtain this test is with its `test` postcommand. Using this postcommand after the model with all the explanatory variables will implicitly compare the two models. Those results can be found in Figure 13.3.

Here, we have asked Stata to jointly test whether the *age* and *rincome06* coefficients are equal to zero (see Figure 13.3). This is the same as asking whether the second model fits the data better than the first model. We used a nested *F* test to compare linear regression models in a similar fashion (see chapter 5). Our results suggest that there is a difference when the values are in the model. Thus, we gain confidence that at least one of these variables should be included in the model.

Another method for comparing logistic regression models is with what are known as information criterions. The two most widely used are called *Akaike's information criterion* (AIC) and the *Bayesian information criterion* (BIC). Both use the deviance value from the estimated models. The standard formulas for these statistics (there are others) are

$$\text{AIC} = \text{Deviance} + (2 \times k)$$

$$\text{BIC} = \text{Deviance} + (\ln(n) \times k)$$

The k in these equations represents the model's degrees of freedom (number of explanatory variables + 1); the n represents the sample size. The rule of thumb for both statistics is that smaller values indicate better fitting models. In some programs, the AIC and BIC are computed using different formulas; however, regardless of the formula used, the rule of thumb remains as small is better.

Table 13.6: Model Fit Statistics

Test	Model	
	Nested	Full
Akaike's information criterion	942.134	898.515
Bayesian information criterion	951.444	917.137

In Stata, we may request these two statistics by using the postcommand `estat ic`. The AICs and BICs for the two logistic regression models of neighborhood fear are provided in Table 13.6. Both statistics indicate that the logistic regression model with gender, age, and income (the full model) provides a better fit to the data than the model with just gender (more information about these and other fit statistics is provided in Zucchini, 2000).

Binary logistic regression provides a powerful tool for evaluating outcome variables that take on only two values. These types of variables are not uncommon in the various scientific disciplines that use statistical modeling. Rather, they are of broad interest and provide important information about many relevant phenomena. We have presented only a brief introduction to this valuable regression technique. Moreover, we have not addressed a twin model: probit regression. Probit regression provides similar results in general (although the coefficients are scaled differently) but is based on the assumption that underlying a binary variable is a continuous variable that is represented along a continuum from a probability of zero to a probability of one. Substantially more information on logistic and probit regression is provided in the numerous books that address categorical outcome variables.

Key Terms

The following is a list of key terms and concepts in chapter 13. You can find definitions to these words throughout the chapter.

- Adjusted odds ratio
- AIC
- BIC
- Deviance
- Linear probability model
- Log-odds
- Logistic function
- Logistic regression
- Maximum likelihood
- Odds
- Odds ratio
- Percentage change formula

Analysis Exercise

1. Create a variable for having a college degree from the *educ* variable that is defined as having more than or equal to 16 years of education. Then estimate a model predicting college degree as a function of age, gender, father's education (*paeduc*), mother's education (*maeduc*), and number of siblings (*sibs*). Interpret the odds ratios for age, father's education, and siblings.

Solution

1. A one-year increase in age increases the odds of a college degree by 3.8 percent. A one-year increase in father's education increases the odds of a college degree by 19.6 percent. A one-sibling increase is associated with a 11.2 percent decline in the odds of a college degree.

CHAPTER 14 AN INTRODUCTION TO MULTILEVEL MODELS

Chapter Objectives

In this chapter, we will briefly introduce models that are created to handle clustered data, or data where observations are not independent. Such models cannot be run in an OLS framework.

In chapter 14, we will

- discuss hierarchical data
- introduce regression that handles clustered data
- briefly introduce the multilevel model and its usefulness
- highlight extensions of the multilevel model.

A number of disciplines in the social, behavioral, and natural sciences regularly use data that are quantitative and *hierarchical.* Although the term *quantitative* is clearly understood, the term *hierarchical* is appreciated less often. Hierarchical data refer to information that is collected at multiple levels of abstraction. Levels of abstraction are often referred to as the "units of analysis," such as individuals, informal social groups, schools, organizations, neighborhoods, or nation-states. Perhaps a couple of examples may help clarify this issue.

One example comes from educational research, where scholars sometimes collect information from students, teachers, and school administrators. Because such data collection is usually expensive and time consuming, researchers typically collect data on samples of students in particular classrooms, the teachers who direct these classrooms, and the administrators to whom these teachers report. Hence, students are *grouped* or *nested* within classrooms, and teachers are grouped or nested within schools. In practice, the researcher has data on individual characteristics from students, classroom characteristics from teachers, and school characteristics from administrators.

A second example comes from clinical research, often conducted by social workers, clinical psychologists, and marriage and family therapists, that addresses the

Table 14.1: Example of Nested Data

idcomm	id	income
afton	10001	2
afton	10004	4
afton	10005	3
afton	10008	2
afton	10009	8
audubon	110069	1
audubon	110070	4
audubon	110071	4
audubon	110074	2
audubon	110076	6

effectiveness of various therapies to treat mental disorders such as depression, anxiety, or phobias. Data are collected on patients who are undergoing specific types of therapies so that these regimens may be compared directly. To make studies efficient, the patients are often drawn from lists of specific therapists, some of whom provide each type of therapeutic regimen. Consider how this is similar to the data collection effort described in the previous paragraph: Patients are nested within types of therapies but also by specific therapists. Any one therapist may, for example, treat several of the patients who are participating in the study. Yet these therapists also have particular characteristics (perhaps some are more empathetic than others) that may affect treatment and, ultimately, the outcome of interest (which may be recovery or relapse, depending on the project).

Our discussion of multilevel models is simply an introduction to the model. Numerous books and courses focus on this method alone. However, we want to provide a few examples of multilevel modeling, by way of introduction. We begin with a discussion of what kind of data are needed to use multilevel models. For a concrete example of hierarchical data, consider Table 14.1. It shows a small selection of data from a study that sampled residents of 99 towns in the state of Iowa (*iowa12.dta* if you are using Stata and *iowa12.sav* if you are using SPSS). The variable *idcomm* references the particular Iowa town, whereas the *id* variable indicates individuals who live in these towns. The variable *income* is a categorical variable that gauges annual household income. It is simple to see that there are multiple survey respondents who are nested within each town.

Unfortunately, using quantitative techniques that ignore the hierarchical nature of data may lead to incorrect inferences and conclusions. As we know from previous

Table 14.2: Ordinary Least Squares Regression of Income

Variable	*b*	*SE*
Male	0.518	0.039***
Married	1.748	0.042***
Intercept	2.931	0.036
N	9,859	
R^2	.181	

***$p < .001$.

chapters, an important assumption of many quantitative models is that the observations are statistically independent; that is, that observations do not influence one another or share unobserved characteristics. Yet there is clear evidence that students grouped in classrooms, to use a well-known example, both influence one another and share characteristics that could affect numerous outcomes. Using an OLS regression model to determine whether a student-level characteristic, such as self-esteem, affects an outcome, such as being disruptive in the class, is inappropriate because students who share a classroom are clearly not independent. The result of using such a statistical model leads to incorrect inferences about the true effect of the explanatory variables on the outcome variable. Thus, any conclusions drawn from this type of statistical exercise will be misleading.

A simple approach to consider when data are hierarchical and the analyst does not care about the potential effects of group-level variables or variability across groups is to identify a variable that gauges the nesting structure, such as the *idcomm* variable shown earlier. This variable is then used to adjust the standard errors in a regression model. Details of this type of approach are provided in many texts that concern survey sampling. In Stata, the nesting variable, or group identifier, is used along with the `cluster` command (use the option `cluster`) to adjust the standard errors for the nonindependence of observations. In SPSS, this is achieved through the drop-downs *Analyze-Classify-Hierarchical Cluster.* In the following example, suppose that we are interested in predicting the income variable shown earlier, using as explanatory variables *male* and *married.* A linear regression model that ignores the nesting by town is shown in Table 14.2.

However, the standard errors in this model are likely biased because of the nonindependence of observations. Clustering may be used to adjust the standard errors for the possible dependence due to sample members who live within the same town. In this case, the standard errors are adjusted for the 99 clusters (towns) identified by *idcomm* variable. Notice that the standard errors are larger in the model (as shown in Table 14.3) that takes into account the hierarchical data structure. This is often the case. However, if there is no clustering effect, the standard errors will be the same in both models.

Table 14.3: Clustered Regression of Income

Male	0.518	0.042***
Married	1.748	0.044***
Intercept	2.931	0.041
N	9,859	
R^2	.181	

***$p < .001$.

Sometimes researchers want to go beyond the individual effects of variables on one another by modeling how effects may vary across groups like schools, organizations, neighborhoods, and so forth. Moreover, researchers are often interested in the effects of group-level variables on individual-level outcomes. Fortunately, recent advances in statistical modeling allow researchers to take advantage of hierarchical data structures. The statistical models designed to analyze these types of data and variables are known as *multilevel models*.

The Basic Approach of Multilevel Models

Multilevel models—also known as *hierarchical models*, *random effects models*, and *mixed effects models*—stem from developments in agricultural statistics, educational statistics, biometry, and econometrics. It was recognized long ago in agricultural studies of crop growth that an efficient design called for growing sets of plants in shared environments. Yet if one wished to extrapolate findings to the broader "population" of plants, one needed to be able to consider that the relationships found in an analysis might be "random." Hence, as inferred earlier, a model that sees these relationships as different, or random, rather than fixed across units of, for example, agricultural plots are called *random effects models*. Not only do the characteristics of a plant vary (because we have many plants to examine), but the relationships between these characteristics (for example, how much water the plant receives) and some outcome of interest (for example, plant growth) vary.

In general, a multilevel model separates the variability of the outcome variable into two components: the variability that is due to differences across individual units (for example, plants, people) and the variability that is due to differences across groups (for example, plots of land, schools, neighborhoods). As an example, consider the *iowa12.dta* data set that includes individual respondents grouped within towns. We will continue to focus on the variable *income*. This variable has an overall mean of 4.36 (recall that it is measured in categories), a standard deviation of 2.11, and a variance of 4.45.

However, think about the towns that the respondents reside in. Each town may have a different mean, reflecting higher or lower average incomes among respondents. Thus, whereas there is overall variability among the individuals in the data set, there is also variability across towns. Another way of stating this is that, although there is an overall fixed mean in the data set, there is also random variation in the means across towns. This is where the term *random effects* comes from.

Examining the variability across individuals and communities is similar to conducting a one-way ANOVA model. Just as we wish to compare variances within units and across units in a one-way ANOVA model, here, we compute and compare the variance that is due to individual-level variation and that is due to community-level variation. We may also set up a statistical model by using the following notation:

$$y_{ij} = \alpha_j$$

In this notation for an *intercept-only model*, the y has subscripts to indicate that there are multiple observations (i), but there are also multiple groups (designated by the j subscript) within which these observations are nested. The α has subscript j because there are multiple means across the groups. In a multilevel model, we estimate three parameters: the overall mean (α_0, or *grand mean*, as it's often called), the between-group variance (σ^2_u), and the within-group variance (σ^2_e). These may also be referred to as *between-group* and *between-individual* variability. However, these terms are dependent on the type of hierarchical data. As shown later in the chapter, with longitudinal data, time periods are nested within individuals so the terms *between-individual* and *within-individual* variance might be used. In Stata, one way to estimate these parameters is by using the `xtmixed` command.

This command requires two things: an outcome variable and an indicator of the variable that designates how the individuals are grouped. As we learned earlier, in the *iowa12* data set, this latter variable is *idcomm*. The following commands provide a one-way ANOVA for the variable *income*. In Stata, we use the `xtmixed` command (in this case, the full command is `xtmixed income || idcomm:`). The double bar separates the command into the *fixed effects* component and the *random effects* component. In this model, the random effects are allowed to vary across communities that are designated by the *idcomm* variable. In SPSS, we can fit this model by using the drop-downs *Analyze-Mixed-Linear* and placing the cluster variable in the box labeled *Subjects*. Both the Stata and SPSS outputs indicate that there are 9,859 individual observations nested within 99 groups (Iowa towns). The minimum number of observations in a town is 70 and the maximum is 119, with a mean of almost 100 per town. Table 14.4 indicates the grand mean (α_0) of *income*, which is about 4.36. This is also known as the *fixed effect* estimate for this model. The *random effects parameter* shows the estimates for the

Table 14.4: Intercept-Only Model

Variable	Coefficient	*SE*
Fixed components		
α_0	4.362	0.047***
Random effects parameter		
Community	0.415	0.037
Residual	2.069	0.015
N	9,859	
Groups	99	
Intraclass correlation coefficient	.039	

***$p < .001$.

variability of *income*, or the random effects portion of the model. As described earlier, there are two estimates. The first is the variation that is due to differences between communities (σ_u). The second is variation that is due to differences within communities (or between individuals) (σ_e). Notice that the latter is larger than the former; this is common in community studies because variability in attitudes, behaviors, and so forth is typically larger across individuals than it is across groups.

One way to characterize this variability is with a ratio measure of the community-level variance to the total variance. For this model, we may compute this as

$$\rho = \frac{\sigma_u^2}{\sigma_u^2 + \sigma_e^2} = \frac{0.415^2}{0.415^2 + 02.069^2} = 0.039$$

This quantity is called the *intraclass correlation coefficient*, or ICC. It measures the proportion of variability in the outcome that is between groups (towns). Thus, about 3.9 percent of the variability in *income* is between towns. The ICC is also called the *cluster effect* or the *intra-unit* correlation.

The ICC is useful because if there is no effect of clustering, then standard regression models may be used. If it is a substantial number, then it is important to take into account the clustering as this suggests that the observations are not independent. Although there is no standard rule of thumb for the size of the ICC, some analysts recommend that it be used to compute the *design effect*, which is the ratio of the sampling variance for the statistic computed using the hierarchical design divided by the sampling variance that would have been obtained assuming a simple random sample. A design effect of two or more is usually indicative that the nesting is substantial enough to warrant a multilevel model (or some other statistical adjustment). The ICC may be combined with the average group size to yield the design effect:

$$d^2 = 1 + (\widehat{group\,size} - 1) \times ICC = 1 + (99.6 - 1) \times 0.039 = 4.85$$

The 99.6 in the equation is the average group size. Notice that the larger the average group size, the larger the design effect. Because the design effect is substantially greater than two, it is best to rely on a multilevel model for this outcome variable. A likelihood-ratio chi-square test may also be used to compare a linear regression model with this model. The null hypothesis is that they are the same, statistically speaking. Here, this test indicates that we may reject the null hypothesis with considerable confidence and conclude that there is a substantial hierarchical effect for this outcome.

Multilevel Regression Model

Let's now consider a hypothesis that suggests a link between income and another variable in the data set. One hypothesis we may wish to consider is whether males have higher incomes than females across towns in Iowa. Hence, we will explore the association between the variables *male* and *income*. Although we already know that mean income varies across communities, our concern can now change. We may now ask whether the regression slopes that indicate the association between *male* and *income* vary across communities.

First, note that a linear model that does not take the hierarchical structure into account is as follows:

$$\widehat{income} = 3.989 + 0.834(male)$$

Thus, with a very small *p* value, we may say that, on average, males report higher annual household incomes than females. Females report, on average, annual incomes of about 3.989 units, but males report incomes about 0.834 units higher. Again, however, this is an overall effect that does not take into account differences across towns.

Using Stata's `xtmixed` command or the SPSS drop-downs we highlighted earlier, we will first estimate a model (shown in Table 14.5) that simply takes into account the nesting of individuals within towns to estimate the variability within and across groups. This is commonly called a *random-intercept model* because the intercept is allowed to vary randomly across the groups.

A more interesting model examines whether the slopes also vary randomly across towns. This extends the model estimated earlier by telling Stata or SPSS that we wish to test a random-slope, random-intercept model. We suggest that you use Stata in this case because the hierarchical modeling menus in SPSS are rather unintuitive. In Stata,

Table 14.5: Random Intercept Model

Variable	Coefficient	*SE*
Fixed components		
Male	0.815	0.041***
α_0	4.362	0.047***
Random effects parameter		
Community	0.402	0.036
Residual	2.029	0.015
N	9,859	
Groups	99	
Intraclass correlation coefficient	.038	

***$p < .001$.

we can run this model with the following command line: `xtmixed income male || idcomm: male,` where the `idcomm: male` portion tells Stata to let the *male* slope vary across towns (data not shown here).

In this model, we test the hypothesis that the slope gauging the linear association between sex and income varies across towns. As shown by the random effects panel, there is variability in the *male* coefficient (data not shown here). A straightforward way of representing this variability is by computing 95 percent confidence intervals for the slope by using the standard formula.

$$95\%\ \text{CI} = 0.814 \pm 1.96(0.129) = [0.561, 1.067]$$

Thus, we may say that we expect about 95 percent of the slopes to fall within the interval 0.561 and 1.067. In other words, in some towns, the slope of the relationship between sex and income (that is, the average income difference between males and females) is much larger than in other towns. But it is still positive across towns in Iowa.

Group-Level Variables and Cross-Level Interactions

A key advantage of multilevel models is that they may include not only individual-level variables, such as sex and marital status, but also group-level variables. For example, in the *iowa12.dta* data set, there are variables that measure the towns' population, economic disadvantage (at the county level), and religious heterogeneity (an index of the number of different religious groups represented in the community). We might be interested in whether, say, household income is lower, on average, in larger towns or in

Table 14.6: Hierarchical Linear Modeling with Predictors at Levels 1 and 2

Variable	Coefficient	*SE*
Fixed components		
Male	0.815	0.043***
Population	0.0001	0.001***
Disadvantage	0.094	0.043*
α_0	3.857	0.061***
Random effects parameter		
Community	0.364	0.035
Male	0.115	0.094
Residual	2.029	0.015
N	9,859	
Groups	99	

*$p < .05$. ***$p < .001$.

more disadvantaged areas. Table 14.6 shows a model that tests this. Consider that we simply add the group-level variables to the fixed portion of the model command.

The results indicate that towns with more people tend to have individuals who report higher household incomes. What is surprising, however, is that those counties that have greater economic disadvantage have sample members from towns within them who report higher household incomes, on average. Although unexpected, this is the sort of statistical association that can be revealed in a multilevel analysis.

Researchers often wish to present some measure of the proportion of variance explained by a regression model. For example, the well-known R^2 statistic is usually presented for a linear regression model when OLS is used to estimate the parameters. In multilevel models, there are analogous measures. Perhaps the most common is variance explained at the lowest level of aggregation (among individuals in this model). This requires information from the within-group variances (σ^2_e) from the one-way ANOVA model and the hypothesized model. For the previously estimated model, this is computed as

$$\frac{\sigma_e^2(ANOVA)-\sigma_e^2(hypothesized\,model)}{\sigma_e^2(ANOVA)}=\frac{2.069^2-2.029^2}{2.069^2}=0.039$$

This indicates that about 3.9 percent of the variance in income within the towns, and hence between individuals, is explained by the model that includes sex, community-level disadvantage, and population size.

Figure 14.1: Stata Results

```
xtmixed income c.age##c.disadvantage || idcomm: age

Mixed-effects REML regression                   Number of obs      =      9859
Group variable: idcomm                          Number of groups   =        99

                                                Obs per group: min =        70
                                                               avg =      99.9
                                                               max =       119

                                                Wald chi2(3)       =   1311.68
Log restricted-likelihood = -20636.299          Prob > chi2        =    0.0000

------------------------------------------------------------------------------
       income |      Coef.   Std. Err.      z    P>|z|     [95% Conf. Interval]
------------------------------------------------------------------------------
          age |  -.0415301   .0011497   -36.12   0.000    -.0437836  -.0392767
 disadvantage |   .2440765   .0780107     3.13   0.002     .0911783   .3969746
      c.age#c. |
 disadvantage |  -.0031144   .0011768    -2.65   0.008    -.0054208  -.000808
        _cons |   6.715485   .0759655    88.40   0.000     6.566595   6.864375
------------------------------------------------------------------------------

------------------------------------------------------------------------------
  Random-effects Parameters  |   Estimate    Std. Err.     [95% Conf. Interval]
-----------------------------+------------------------------------------------
idcomm: Independent          |
                     sd(age) |   1.04e-07    .0000248     7.8e-212  1.4e + 197
                   sd(_cons) |   .3355354    .0324331     .2776263   .4055236
-----------------------------+------------------------------------------------
                sd(Residual) |   1.946153    .0139316     1.919039   1.973651
------------------------------------------------------------------------------
LR test vs. linear regression:   chi2(2) =  146.61   Prob > chi2 = 0.0000
```

We may also estimate the proportion of variance explained between groups by using information from the between-group variances (σ^2_u) in the one-way ANOVA model and the hypothesized model. For the previous model this is

$$\frac{0.415^2 - 0.364^2}{0.415^2} = 0.23$$

Thus, about 23 percent of the variability in income between towns is accounted for by the hypothesized model. Given that the male slope also has a variance component, we could also determine the proportion of variance explained between groups that is due to the random slope component.

Another advantage of multilevel models is their ability to test interesting hypotheses about contextual effects on individual-level associations. Using what are termed *cross-level interactions*, we may determine, for example, whether the association between sex and income or between age and income differs depending on characteristics of the town. Figure 14.1 examines whether the association between age and income depends on whether the surrounding county is high or low on economic disadvantage. We have run it in Stata for ease of interpretation, because it is much more intuitive than in SPSS. The results suggest that age is negatively associated with income (perhaps because of the older average age of this sample) but that this negative association is slightly stronger in counties with higher levels of economic disadvantage (see Figure 14.1). It is often simpler to interpret these effects if the variables are first centered and then included in the model. There is a substantial and important body of literature on the use of centering in multilevel models. One of the key issues is that one may center individual-level variables within group or overall. The type of centering has implications for the interpretation of model results. Books on multilevel modeling, such as Heck and Thomas (2008), provide more information.

Extensions

What we have provided here is a very simple introduction to multilevel models. In one sense, many researchers care only about accounting for the clustered nature of their data and are less concerned about explicitly modeling the effect of variables at the individual level and at a higher level. In this case, running regular regression models, as we have shown here, with standard error corrections for the clustered nature of the data may be enough. In other cases, you may want to place macro-level predictors into the model, which leads to more complicated analyses. In general, you will want to test the assumptions of all models when you run them. We strongly recommend that if you are interested in running multilevel models, you consult one of the many top-notch books on these models available in the market.

Hierarchical modeling has many extensions. For example, a researcher can run some longitudinal models in a hierarchical framework. One popular kind of model is the growth curve model. Many researchers tend to use structural equation modeling to run growth curve models, which assess changes in an outcome over a period of time and look at the influence of explanatory variables on inflections in the curve. However, these models are more easily run in a hierarchical linear modeling framework where individual data points are nested within individuals themselves. One reason such modeling is easier is because of less restrictive assumptions in hierarchical modeling, in addition to easier specification of the model.

A second popular extension, which we alluded to in this chapter, is to use models that consider how personal and structural characteristics influence one another to produce individual outcomes. If you think about it, such models, although rarely seen in social work, capture the social work "model" better than a standard regression model in many ways. In particular, social workers acknowledge that human behavior is rooted in biopsychosocial processes. As such, individual behavior is a process of individual factors rooted in environmental and cultural contexts. Considering this, perhaps social work researchers should more actively pursue modeling and research questions that require hierarchical data.

As we have mentioned several times, we wanted to provide only a brief introduction to hierarchical or multilevel modeling in this chapter. It seems that as social work research and other social sciences advance, hierarchical data will become more and more commonplace. For us, this is a double-edged sword. On the one hand, it means that we will have significantly better data that will improve our understanding of the complex processes that govern human behavior. On the other hand, such data will require increasingly sophisticated modeling and a better understanding of the common statistical models used in social science research. Thankfully, the regression model, which we covered in substantial detail here, forms the backbone of most statistical analyses—including the multilevel model.

Key Terms

The following is a list of key terms and concepts in chapter 14. You can find definitions to these words throughout the chapter.

- Cross-level interaction
- Fixed effects model
- Grand mean
- Group/nested/clustered
- Hierarchical data
- Intercept-only model
- Multilevel models
- Random effects model
- Random-intercept model

CHAPTER 15 CONCLUSION

The linear regression model offers a powerful tool for analyzing the association between one or more explanatory variables and a single outcome variable. Some novice researchers wish to move quickly beyond this model and learn to use more sophisticated models because they get discouraged about its limitations and believe that other regression models are more appropriate for their analysis needs. Although there are many situations where alternative regression models are needed (see, for example, chapter 13), the linear model should not be overlooked or underused. Even when one or more of the assumptions described in the previous chapters is violated, linear regression will frequently provide accurate results. Although saying this model is robust is ill advised because *robustness* has a very specific statistical meaning, it is not disingenuous to claim that the model works well under a variety of circumstances. There are myriad practical uses in the social, behavioral, and health sciences for linear regression analysis. Furthermore, the linear model is truly the building block of other, more advanced statistical methods.

The key to using linear regression analysis appropriately is to understand its strengths and weaknesses. It works quite nicely with dummy and continuous explanatory variables. As long as the outcome variable can be transformed to something close to normality, the coefficients and standard errors are within a reasonable distance of the true population parameters (assuming, of course, that we have a good sample). Also with adjustments to the standard errors that minimize the negative effects of heteroscedasticity and influential observations, the results can move even closer to the precision most of us desire. In particular, Huber-White-Sandwich estimators are especially useful when heteroscedasticity or influential observations show up or are suspected (see chapters 11 and 12).

However, linear regression models should not be used when the outcome variable is binary or takes on only a few values. How many is "a few" has been a point of contention in the research literature. Some claim less than nine, others less than seven. The key, though, is whether the variable is distributed normally, not precisely how many categories it has. Nevertheless, given the availability of the many regression models designed for categorical outcome variables, there is little need to rely on linear regression analysis in this situation. The binary logistic regression model, for example, provides a solid alternative when the outcome variable is binary. Other models are described in the many books on categorical data analysis and generalized linear models.

There are many other topics that we have not had time to address in this presentation. For example, the issue of nonindependence was mentioned only briefly when discussing the assumptions of linear regression analysis (chapter 2) and autocorrelation (chapter 11). Nonindependence occurs when we have longitudinal data or spatial data but also in other contexts. Recall that standard errors are biased in the presence of nonindependent observations.

The general issue of nonindependence of observations is particularly germane to the more specific topic of survey sampling. It is rare to find large surveys that do not use some form of sampling that leads to nonindependence of observations. These sampling schemes are conducted for efficiency and cost-effectiveness. It is prohibitively expensive to gather a simple random sample of people in the United States. Hence, most national surveys (and many state surveys) use some form of *stratified* or *cluster* sampling. This means that units such as metropolitan areas are sampled first, followed by smaller units such as census tracts, block groups, and households. There are usually several stages to the sampling. Telephone surveys typically use groups of exchanges that also result in a certain degree of clustering. When using most national data sets, it is, therefore, important to consider the effects of stratification and clustering on regression estimates. As mentioned earlier, the main effect is on the standard errors, so they should be adjusted. Many statistical software packages now include techniques for survey data that adjust standard errors for nonindependence. For example, Stata includes a set of statistical techniques for survey data that include nonindependent observations. Type `help survey` to read about the many options. Furthermore, SPSS has introduced a module called *Complex Samples* that allows analysts to consider nonindependent observations from surveys.

Another issue related to sampling involves the use of weights. *Sampling weights* are used to designate the number of people (or other units) in the population each sample member represents. Recall that we want our sample to represent some well-understood population. Therefore, each observation in our sample should represent some group from the population. For example, in a nationally representative sample, a 40-year-old Caucasian man may represent several thousand 40-year-old Caucasian men from some part of the United States. Survey data often include sample weights that may be used in analyses (for example, our 40-year-old man has a weight of 12,340 because he represents 12,430 40-year-old men). Yet if these weights are whole numbers designed to indicate the actual number of people each observation represents, the presumed sample size can be inflated tremendously. What effect do larger samples have on standard errors? All else being equal, they make them smaller. Experience shows that almost any regression coefficient is statistically significant—whether or not it is important—if the sample is weighted to represent a large population. This may be seen as an advantage, but it can also be misleading and give us dubious results. One solution is to compute *normed sampling weights* for the analysis. These are computed as:

$$\bar{w}_i = \frac{w_i}{\bar{w}}$$

In words, take each observed weight and divide it by the overall mean of the weights.

Let's say our national sample has a mean weight of 15,000 (on average, each observation represents 15,000 people). Our 40-year-old man will then have a normed sampling weight of 12,430/15,000 = 0.829. If another observation represents more people, say 17,500, then its weight is 17,500/15,000 = 1.167. Taking the normed weights not only makes the weights smaller and more manageable but also preserves their utility because some observations still have more influence than others. It also results in more reasonable standard errors.

There are many other issues we could discuss, but space is limited. The information provided in the previous chapters should provide the groundwork for further coursework on and self-study of regression modeling. Thus, perhaps it is best to ask interested readers to pursue additional topics on your own through the many excellent books and courses on linear regression and related statistical techniques. Topics such as sample selection bias, nonlinear models, bootstrapping methods, multilevel linear models, generalized linear models, simultaneous equations, methods of moments estimators, generalized additive models, marginal models, event history analysis, and log-linear models (including transition models) have all been used to complement, supplement, or replace linear regression models in one way or another.

Summary

To summarize what we have learned, here is a list of steps to consider as you pursue linear regression analyses on your own. It is not necessarily a comprehensive list as each analysis and model has its own idiosyncrasies that make it unique. Nonetheless, it prescribes the most common steps that should normally be taken as you estimate linear regression models.

How to Conduct a Linear Regression Analysis: A Summary

1. **Know the background literature on your subject.** Does previous literature indicate that your outcome variable is not distributed normally or is not continuous? Has previous literature found interactions between explanatory variables? What control variables are important for your model? Do you anticipate

measurement error? What might you do to correct for it? What theoretical frameworks or conceptual models have been used in the past? Use your imagination and intuition about the social or behavioral processes of interest.

2. **Establish hypotheses and decision rules.** Which theory or conceptual model guides your selection of variables or the ways you think they are associated? Why do you think the variables are associated? Will you use directional or nondirectional hypotheses? Will you use one- or two-tailed significance tests to decide whether coefficients are statistically distinct from zero? Set up the hypotheses.
3. **Know the data set.** Are your data longitudinal? If they are, consider using a regression technique—such as a generalized estimating equation (GEE)—that is designed for longitudinal data. Do your data come from a survey? Have you read the codebook and documentation for information about how the data were collected? How are the variables of interest coded? Are there missing values? How are they coded? How will you handle missing data? You will probably need to recode and construct new variables. Are there sample weights in the data set? Will you use a split-sample approach to verify the model?
4. **Check the distributions of the variables.** Use *q*-*q* plots and box-and-whisker plots. Run descriptive statistics. In particular, are the explanatory variables normally distributed? If they are not distributed normally, consider using a transformation such as the natural logarithm, square root, or quadratic (depending on the direction and degree of skewness). If the outcome variable is binary, use logistic regression. Are there outliers or other influential observations that you can see? Consider their source. Do you need to compute dummy variables and include them in your model? Check to make sure you've taken care of missing values—they can throw everything off if they are not adjusted for in the model.
5. **Assess the bivariate associations in the data.** Use scatter plots for continuous variables. Plot linear and nonlinear lines to determine the bivariate associations. Compute bivariate correlations for continuous variables. Look for outliers and potential collinearity problems.
6. **Estimate the regression model.** Avoid automated variable selection procedures unless your goal is simply to find the best prediction model. Assess the results. Are there unusual coefficients (overly large or small; negative when they should be positive)? Save the collinearity diagnostics; save the influential observation diagnostics (studentized residuals, leverage values, Cook's *D*, and DFFITS); run a scatter plot of deleted studentized residuals by standardized predicted values; estimate partial residual plots; ask for a normal probability plot of the residuals; ask for the Durbin-Watson statistic, if appropriate; compute Moran's *I*, if using spatial data and it is available. Assess the goodness-of-fit statistics (adjusted R^2;

F value and its accompanying *p* value; S_E). Run nested models, if appropriate, and use nested *F* tests to compare them. These are particularly useful for assessing potential specification errors.

7. **Check the diagnostics.** Are there any collinearity problems? If yes, you might need to combine variables, collect more data, or, as last resort, drop variables. If the collinearity problem involves interaction or quadratic terms, use centered values such as z scores to recompute them. Are the residuals normally distributed? If not, consider a transformation. Do the partial residual plots provide evidence of nonlinear associations? Is there evidence of heteroscedasticity? (If the plot is inconclusive visually, try White's test.) If yes, consider transforming a variable, using weighted least squares regression, or using Huber-White-Sandwich estimators. Is there evidence of autocorrelation? Consider the source and try to correct for it. If you have spatial data, a spatial regression model may be needed. Use Prais-Winsten regression or time-series techniques for data collected over time, if appropriate. Are there influential observations? If yes, consider their source. Are there coding errors? Will a transformation help? If not, use a robust regression technique to adjust for influential observations.
8. **Interpret and present the results.** Interpret the unstandardized slopes and *p* values. What do the goodness-of-fit statistics tell you about the model? Compare the results with the guiding hypotheses. Given the decision rules, are the hypotheses or the conceptual model supported by the analysis? Consider presenting predicted values, especially from models that include interaction terms. Consider graphical presentations of coefficients, nonlinear associations, and interactions. These can provide intuitive information that is often lost when presenting only numbers.

REFERENCES

Aiken, L. S., & West, S. G. (1991). *Multiple regression: Testing and interpreting interactions.* Thousand Oaks, CA: Sage Publications.

Bartholomew, D. J. (1987). *Latent variable models and factor analysis.* New York: Oxford University Press.

Beider, P. P., Groves, R. M., Lyberg, L. E., Mathiowetz, N. A., & Sudman, S. (Eds.). (2004). *Measurement errors in surveys.* New York: John Wiley & Sons.

Bradburn, N. M., Sudman, S., & Wansink, B. (2004). *Asking questions: The definitive guide to questionnaire design—For market research, political polls, and social and health questionnaires.* San Francisco: John Wiley & Sons.

Cleveland, W. S. (1993). *Visualizing data.* Summit, NJ: Hobart Press.

Davidson, R., & Mackinnon, J. G. (1993). *Estimation and inference in econometrics.* New York: Oxford University Press.

Draper, N. R., & Smith, H. (1998). *Applied regression analysis* (3rd ed.). New York: John Wiley & Sons.

Ervin, L. H. (2000). Using heteroscedasticity consistent standard errors in the linear regression model. *American Statistician, 54,* 217–224.

Fox, J. (1997). *Applied regression analysis, linear models, and related methods.* New York: Sage Publications.

Graybill, F. A., & Iyer, H. K. (1994). *Regression analysis: Concepts and applications.* Belmont, CA: Duxbury Press.

Hamilton, L. C. (1992). *Regression with graphics: A second course in applied statistics.* Belmont, CA: Duxbury Press.

Hardin, J. W., & Hilbe, J. M. (2003). *Generalized estimating equations.* Boca Raton, FL: Chapman & Hall/CRC.

Heck, R. H., & Thomas, S. L. (2008). *An introduction to multilevel modeling techniques* (2nd ed.). London: Routledge.

Hoffmann, J. P. (2004). *Generalized linear models: An applied approach.* Boston: Addison-Wesley Longman.

Huff, D. (1954). *How to lie with statistics.* New York: W. W. Norton.

Kleinbaum, D. G., Kupper, L., Azhar, N., & Rosenberg, E. S. (1998). *Applied regression analysis and other multivariable methods* (3rd ed.). Pacific Grove, CA: Duxbury Press.

Kline, P. (1999). *An easy guide to factor analysis.* London: Routledge.

Krosnick, J. A., & Fabrigar, L. R. (2005). *Questionnaire design for attitude measurement in social and psychological research.* New York: Oxford University Press.

Ostrom, C. W. (1990). *Time series analysis: Regression techniques* (2nd ed.). Newbury Park, CA: Sage Publications.

Pearl, J. (2000). *Causality: Modeling, reasoning, and inference.* New York: Cambridge University Press.

Rogerson, P. A. (2006). *Statistical methods for geography* (2nd ed.). Thousand Oaks, CA: Sage Publications.

Salsburg, D. (2001). *The lady tasting tea.* New York: Owl Books.

Schott, J. R. (1997). *Matrix analysis for statistics.* New York: Wiley.

Singer, J. D., & Willett, J. B. (2003). *Applied longitudinal data analysis.* New York: Oxford University Press.

Thompson, B. (2004). *Exploratory and confirmatory factor analysis: Understanding concepts and applications.* Washington, DC: American Psychological Association.

Weiss, N. A. (1999). *Introductory statistics* (5th ed.). Reading, MA: Addison-Wesley.

Zucchini, W. (2000). An introduction to model selection. *Journal of Mathematical Psychology, 44,* 41–61.

INDEX